ANIMAL SCIENCE, ISSUES AND PROFESSIONS

MONKEYS: BIOLOGY, BEHAVIOR AND DISORDERS

ANIMAL SCIENCE, ISSUES AND PROFESSIONS

Additional books in this series can be found on Nova's website
under the Series tab.

Additional E-books in this series can be found on Nova's website
under the E-books tab.

ANIMAL SCIENCE, ISSUES AND PROFESSIONS

MONKEYS: BIOLOGY, BEHAVIOR AND DISORDERS

RACHEL M. WILLIAMS
EDITOR

Nova Science Publishers, Inc.
New York

Library of Congress Cataloging-in-Publication Data

Monkeys : biology, behavior, and disorders / editors, Rachel M. Williams.
p. cm.
Includes bibliographical references and index.
ISBN 978-1-61209-911-8 (hardcover : alk. paper)
1. Monkeys. 2. Monkeys--Behavior. 3. Monkeys--Diseases. I. Williams, Rachel M.
QL737.P9M596 2011
599.8--dc22

2011004584

Published by Nova Science Publishers, Inc. ✝ New York

CONTENTS

PREFACE

This new book focuses on the biology, behaviors and disorders of monkeys. Topics discussed include the use of nonhuman primates in biogerontology; cognitive correlates of communication in primates; effects of the adverse rearing experience on the organization of the brain and behavior among nonhuman primates; parent-infant relationships in Marmosets; planning abilities of monkeys; neuropeptides in the monkey brainstem and developmental neuronal toxicity and the Rhesus monkey

Chapter 1 - Progress in understanding the etiology of human disease and in the development of effective therapies is heavily dependent on appropriate animal models in which to study the underlying causal mechanisms. Because nonhuman primates (NHPs) share many genetic and physiological similarities with humans, they have served as pragmatic models for a wide range of human diseases for several decades. For instance, female rhesus macaques (*Macaca mulatta*), like women, show ~28-day menstrual cycles and eventually undergo menopause. In addition, they show many of the same age-related changes in physiological and behavioral functions, including perturbed sleep-wake cycles and cognitive decline. Additional features make NHPs ideal experimental models for aging research. They can be maintained in carefully controlled environments (e.g., photoperiod, temperature, diet, and medication), and self-selection bias that is often unavoidable in human clinical trials can be completely eliminated with NHP studies. The use of NHPs is also advantageous because similar techniques (e.g., activity recording and MRI) can be used to make comparisons between NHPs and humans. In addition, NHPs are out-bred, which enables rigorous validation of research findings that goes beyond the proof of principle provided by rodent models. Finally, because macaques are long-lived species, it is likely that these NHPs have adapted similar maintenance strategies as humans, making them an ideal translational model. Thus, in this chapter the authors focus on age-related changes in the NHP endocrine, cognitive and immunologic systems, and their contributions to translational research.

Chapter 2 – Early adverse rearing experiments have been performed on nonhuman primates for over 60 years. These studies have generated important insights into the effects that the early social environment can exert on the developmental course of behavioral and physiological systems. Although a number of different models of adverse rearing have been used they all interrupt the interaction between infant and mother and result in abnormalities in similar domains. These include motor development, social behavior, emotional responsiveness, neuroendocrine dysregulation, alterations in central monoamne function, and abnormal development of some brain structures. The authors review these findings in this chapter.

Chapter 3 - Although paternal behavior is rare in mammals including primates, males take an active role in rearing infants among callitrichid species. The repertoire of male–infant interactions observed in these species is similar to that observed in humans, such as carrying, protecting, food sharing, grooming, playing, and proximity. Callitrichid species could be a suitable model for research on paternal behavior as well as maternal behavior and the relationship between parents and infants. In this review, the authors first survey developmental changes in the relationship between infants and parents. Next, the authors introduce experiments in which attachment behavior from infants to parents and from parents to infants was investigated and examine whether the behavior differs between fathers and mothers. In addition, the authors refer to a unique behavior of callitrichid species, namely, food transfer from parents to offspring. Finally, the authors examine how behavior can be controlled via endocrinological variables.

Chapter 4 - Prompted by the theoretical prediction that damage to the hippocampal system should abolish exploratory behaviour, the present study examined exploratory movements in control monkeys (CON) and monkeys with transection of the fornix (FNX), a major input/output pathway of the hippocampus. CON and FNX monkeys were introduced to a large novel octagonal chamber (approximately 7.4 m^2) for six daily sessions each lasting 20 minutes. Both groups visited, punctuated by stops, the majority of the floor-space of the environment in each of the sessions. The exploratory movements of CON and FNX groups were not significantly different on most of the measures taken over 6 consecutive days. These measures included cumulative distance traveled, number and duration of stops, travelling patterns, and proportion of time spent in each of 12 designated zones of floor-space. The high degree of similarity in behaviour between CON and FNX groups suggests that an intact hippocampal system is not necessary for the display of normal exploratory movement per se. On the other hand, the CON and FNX groups did behave differently on two measures. First, the CON group exhibited a decrement in distance traversed over consecutive epochs within the first test session whereas FNX animals did not. Second, on those days in which the chamber was made visually asymmetrical, the CON animals tended to show a predilection for spending proportionally more time within one particular quadrant of the chamber. These observations are consistent with the idea that interrupting normal hippocampal system function by means of fornix transection is detrimental to learning about the spatial layout of environments. As the first attempt to compare exploratory and ambulatory behaviours of monkeys with and without fornical damage in a large open space, I argue that while monkeys with fornix transection still display intact locomotor and exploratory behaviour patterns, their new learning of visuospatial context is impeded.

Chapter 5 - Animals in experiments are traditionally grouped by experimental treatment. Although this is a valuable way to differentiate the groups, alternatively, groups can be distinguished based on cognitive performance. Performance based analysis can yield valuable insights, corresponding to behavior and/or cognition, that might not otherwise be observed. As an example of such an analysis, the authors discuss a cohort of elderly female rhesus macaques who participated in a spatial food port maze navigational test. Circadian activity and pharmacological MRI (phMRI) were assessed in these monkeys *in vivo* and radioligand binding was assessed in post-mortem tissue. Based on cognitive performance in the spatial maze, the cohort of monkeys was divided into Good Spatial Performers (GSP) and Poor

Spatial Performers (PSP). GSP animals were more active during the day and less active at night compared to PSP animals. In addition, GSP animals had a higher percentage change in blood-oxygen-level-dependent (BOLD) signal after a scopolamine challenge, a non-specific muscarinic receptor antagonist, compared to PSP animals. Post-mortem radioligand experiments demonstrated that hippocampal muscarinic type 1 (M_1) maximum receptor binding and receptor binding affinity, hypothesized to have an integral role in spatial learning and memory, was significantly greater in the hippocampus of GSP than PSP animals. In contrast to the hippocampus, M_1 receptor binding was not significantly different in the prefrontal cortex (PFC). Muscarinic type 2 (M_2) maximum receptor binding and receptor binding affinity were not significantly different between the two groups in either brain region. Finally, there were positive correlations between circadian measures and the percentage change in BOLD signal following the scopolamine challenge, as well as M_1 receptor binding measures. Thus, GSP animals sleep more and have anenhanced M_1 receptor function. These data demonstrate the close relationship between BOLD signal changes, circadian activity, and M_1 receptor binding parameters. Distinguishing groups based on cognitive or behavioral performance is valuable for studying neurobiological correlates of performance in nonhuman primates.

Chapter 6 – As adults we plan things we will do today, tomorrow, next week, next month, and even years from now. Our ability to plan is adaptive in the sense that it allows us to anticipate events that may occur in the future and appropriately prepare for them in the present. For example, it is common for young adults to save for their retirement, an event that may be 30 to 40 years in their future. Although children do not plan on time scales as large as this, they still quickly learn to save their pocket money to buy the latest toy or plan their activities for the coming weekend.

Chapter 7 - Monkeys are widely used in the laboratory as an experimental animal model in order to answer several scientific questions related to neuroanatomy, neurophysiology, neuropharmacology, neurology and behavior. Brainstem is the region in which the regulation of reflexes and "unconscious" mechanisms (pain transmission, cardiovascular, respiratory…) is located. These mechanisms are mediated by chemical substances such as neuropeptides. These substances, which show a widespread distribution in both the central and peripheral nervous systems, are neuroactive substances that acting as neurotransmitters, neuromodulators (paracrine and autocrine actions) and neurohormones are involved in numerous physiological actions. In the last thirty years, the knowledge on the distribution and functions of the neuropeptides has increased notably in the mammalian central nervous system. Thus, the authors aim here is to review currently available morphological and physiological data on neuropeptides in the monkey brainstem. The authors shall thus discuss the following aspects: 1) The distribution of the neuropeptides in the monkey brainstem; 2) The anatomical relationship among the different neuropeptides in the monkey brainstem; 3) The coexistence of neuropeptides in the monkey brainstem; 4) The peptidergic pathways in the monkey brainstem; 5) The physiological functions of the neuropeptides in the monkey brainstem.

Chapter 8 – The rhesus monkey, *Macaca mulatta,* is an animal model used to inform aspects of human physiology, pathology, pharmacology, toxicology, and systems biology. Because of obvious limitations it is not possible to thoroughly explore the effects of pediatric anesthetic agents on neurons in human infants or children, nor is it possible to determine the dose-response or time-course for potential anesthetic-induced neuronal cell death in humans.

Due to the complexity of the primate brain, the monkey is often the animal model of choice for neurotoxicological experiments and it is in the rhesus monkey that the phenomenon of interest (anesthetic-induced neuronal cell death in the brain) has been previously reported.

Chapter 9 - The monkey visual system begins with the eyes and includes multiple brain areas and processing streams. Consequently, more of the monkey brain is devoted to vision than to any other sensory modality. Like other sensory systems, the visual system is organized hierarchically both in structure and function, and subsequent levels of the hierarchy are more specialized in their processing of visual features than their predecessors. Visual information is translated from the physical world into a neural code in the retina. This information, contained in the spiking activity of retinal ganglion cells, is then transmitted to the lateral geniculate nucleus (LGN) of the thalamus, a processing station that subsequently supplies the primary visual cortex (V1) with the majority of its input. From V1, visual information is routed into multiple branching streams that encompass more than 20 cortical areas, spanning regions whose response properties range from simple edge detection to complex representations of familiar conspecifics. The study of monkey vision thus requires a vast array of strategies and techniques, from the biochemical analysis of phototransduction in the retina, to computational modeling of neural networks that are capable of the processing feats required to construct a neural representation of our dynamically changing visual world. While the enormity of vision science can be daunting, the authors present in this chapter a basic foundation necessary for understanding how visual information is encoded and processed in the monkey nervous system, giving particular emphasis to the inseparable relationship between anatomy and function. This chapter will emphasize visual processing in the macaque monkey—an old world monkey that has been studied more comprehensively than other primates and serves as a model for understanding primate vision.

In: Monkeys: Biology, Behavior and Disorders
Editor: Rachel M. Williams, pp 1-53
ISBN: 978-1-61209-911-8

Chapter 1

INTEGRATIVE MODELS OF AGING IN THE NONHUMAN PRIMATE

Ilhem Messaoudi[1,2,*], Henryk F. Urbanski[3,4] and Steven G. Kohama[3]

[1]Division of Pathobiology and Immunology
[2]Vaccine and Gene Therapy Institute
[3]Division of Neuroscience, Oregon National Primate Research Center
505 NW 185th Avenue, Beaverton, OR 97006, USA
[4]Department of Behavioral Neuroscience and Department of Physiology and Pharmacology, Oregon Health and Science University, 3181 SW Sam Jackson Park Road, Portland, OR 97239, USA

ABSTRACT

Progress in understanding the etiology of human disease and in the development of effective therapies is heavily dependent on appropriate animal models in which to study the underlying causal mechanisms. Because nonhuman primates (NHPs) share many genetic and physiological similarities with humans, they have served as pragmatic models for a wide range of human diseases for several decades. For instance, female rhesus macaques (*Macaca mulatta*), like women, show ~28-day menstrual cycles and eventually undergo menopause. In addition, they show many of the same age-related changes in physiological and behavioral functions, including perturbed sleep-wake cycles and cognitive decline. Additional features make NHPs ideal experimental models for aging research. They can be maintained in carefully controlled environments (e.g., photoperiod, temperature, diet, and medication), and self-selection bias that is often unavoidable in human clinical trials can be completely eliminated with NHP studies. The use of NHPs is also advantageous because similar techniques (e.g., activity recording and MRI) can be used to make comparisons between NHPs and humans. In addition, NHPs are out-bred, which enables rigorous validation of research findings that goes beyond the proof of principle provided by rodent models. Finally, because macaques are long-lived species, it is likely that these NHPs have adapted similar maintenance strategies as humans, making them an ideal translational model. Thus, in this chapter we focus on age-related changes

[*]E-mail: messaoud@ohsu.edu

in the NHP endocrine, cognitive and immunologic systems, and their contributions to translational research.

I. Introduction

Evolution of the Aged Population

In 2010, the world's population is estimated to be about 6.85 billion people, of which 7.8% are 65 years of age and over [1]. This dramatic increase has occurred despite dire predictions of catastrophic famine dating back to the start of the industrial revolution. What 19th century demographers, such as Malthus, did not predict, was that along with population growth, technical advances in agriculture (fertilization, crop rotation, the "Green Revolution") would mitigate the limiting factor of food supply [2]. Current models show a continued population expansion, with an estimated total of 9.3 billion individuals expected by 2050. Interestingly, the percentage of people at that time who will be >65 years old will continue to grow in proportion to 16.7% of the total world-wide population [1]. This increase in the proportion of older individuals is a source of much consternation, as the predicted societal costs for the care of the enlarging post-retirement population will be a continuing challenge.

Because the socioeconomic structure of many countries is quite diverse, the proportion of elderly in industrialized countries today, such as the U.S., are already close to those future projected world totals, with some 13% of the 2010 U.S. population already over the age of 64 years of age [3].

Early in the 20th century, this modification of population size, growth and the proportion of elderly was recognized and described in the model of Demographic Transition [4]. In this model, countries were classified by their rate of population increase, which illustrated the progression of societies from those with high birth and death rates to that of industrialized countries with low birth and death rates. With increased technological advances in medicine, for example, the treatment of infectious diseases, mortality rates receded, especially in the ranks of the young. Since then, an additional "mortality transition" has occurred, with further reductions/delays in degenerative and man-made diseases in the elderly [5]. Coupled with the post-World War II Baby Boomer expansion and lower birth rates, the U.S. is in the midst of an unprecedented increase in the number of elderly.

The Use of NHPs in Biogerontology

Most of our understanding of the biological changes that occur with aging has come from studies with rodents which offer the distinct advantages of an extensive set of reagents, the presence of genetically modified strains, and a short lifespan that allows for longevity studies of short duration. However, there are several fundamental differences between rodent and human physiology. NHPs, on the other hand, share significant genetic homology as well as physiological and behavioral characteristics with humans.

Therefore, NHPs have emerged as a leading translational model to study various aspects of human aging by offering a unique opportunity to carry out mechanistic studies in a species that closely mimics human biology under controlled conditions. NHPs have already provided irreplaceable models for research areas in which their close evolutionary relationship to humans ensures high fidelity models with predictive value for human diseases. For instance, co-morbidity patterns in aging monkeys closely mirror those seen in humans including the development of age-related diseases such as cataracts, age-related macular degeneration, decline of hearing, loss of motor skills, impaired memory acquisition, diabetes, hypertension, osteoporosis, pancreatic and neurologic amyloid deposition, and atherosclerosis [6-9]. These diseases only appear in rodents with genetic manipulation and do not always recapitulate the clinical picture. In contrast, the appearance of these diseases in NHPs, such as atherosclerosis, is increased with age and with the consumption of a Western diet, as described for humans [6-9]. Furthermore, NHPs are susceptible to either human pathogens or simian pathogens that bear significant homology to human infectious agents [10, 11]. Consequently, NHPs reproduce characteristics and functional sequelae of diseases seen in humans. Moreover, analysis of pharmacokinetics of several drugs in different animal models has shown that NHPs are consistently more predictive of the human pharmacokinetics than rodent models [12]. Thus, the use of aged NHPs to both understand and test innovative solutions to promote healthy aging is increasing.

NHPs used in biomedical research can be classified into two broad groups: Old World monkeys (e.g., macaques) and New World monkeys (e.g., marmosets and squirrel monkeys). The New World monkeys are more phylogenetically distant from *Homo sapiens,* and it is estimated that New World monkeys diverged over 30 million years ago whereas Old World monkeys diverged 15 million years ago [10].

Moreover, in contrast to Old World monkeys, very few reagents have been validated for cross-reactivity with New World monkey cells. Consequently, macaques represent the preferred monkey resource for biomedical research (>70%), and have served as invaluable models for human infectious diseases [13]. However, New World monkeys present some advantages such as smaller size, ease of handling and reduced risk of zoonotic disease transmission, which may result in increased use of these species in the future [14].

Phylogeny and Life History of Macaques

The rhesus macaque (*Macaca mulatta*) genome shares some 93.54% homology to humans at the nucleotide level [15]. In addition to being an out-bred species, the rhesus macaque is also long-lived, suggesting that similar patterns of aging may have evolved in these closely related primate species. In terms of documented longevity, survival times in semi-natural conditions have been reported [16], but environmental issues, such as infection, severely shortened longevity [17].

Animals in captivity have better odds of survival with advanced husbandry in place, and early observations of rhesus macaque colonies indicated a median survival time of about 16 and 18 years for males and females, respectively [18]. Animals in this colony were categorized as young adults (<15 years old), middle-aged (15-20 years old), old (20-25 years old), and senile (>25 years old), with the oldest animal living until 35 years old. A later study in another captive rhesus colony observed 50% survivorship at about 24 years for males and 20 years for females [19].

In another long-term study, rhesus macaques had an average lifespan of 25 years and a maximum lifespan of 40 years [20]. This is similar to a recent publication where the 50%

survivorship of controls in a dietary restriction study was a little over 25 years [21]. Therefore, for rhesus macaques in captivity, current estimates places the average lifespan at 25 years and a species maximum of about 40 years, with old animals being in their late teens-early 20s. Because the oldest human survived to over 120 years, rhesus macaques have a maximum lifespan approximately one third that of humans.

Advantages and Disadvantages of NHPs as Aging Models

As with all other experimental animal models, NHPs present several advantages and challenges. Major strengths include their genetic similarity to humans, controlled experimental conditions and the reduction of confounding factors in a manner superior to clinical research [22].

Another advantage of NHPs is their larger size, which allows for longitudinal and cross-sectional assessments of multiple organ systems. These advantages provide the possibility of testing new drugs, vaccines and interventions in a truly translation model. However, compared to other laboratory animals the lifespan of NHPs is relatively long, which itself presents challenges when using NHPs in bio-gerontology research. While a 25-year lifespan is considerably shorter than that of humans, it is still prohibitive for longitudinal studies.

Moreover, limited availability of aged NHPs and the high costs associated with NHP research and the specialized housing requirements can be prohibitive for most researchers. Finally, the ever-increasing level of regulation concerning the use of NHPs in research can be a major burden. To overcome some of these limitations, the National Institute on Aging (NIA) has developed specialized resources to encourage and promote the use of NHP models in aging research, as discussed below.

Resources and Colonies

The National Institutes of Health (NIH) support eight National Primate Research Centers (NPRCs) that provide an effective infrastructure to supply and house NHP for the benefit of research into human diseases. Macaques represent the majority of the NHPs housed at the NPRCs, but other old world species such as African green, sooty mangabey and baboons as well as several New World species such as marmosets are also housed at the NPRCs. The NIA has made significant investments in resources to facilitate the use of NHPs in aging research. These include the maintenance of several colonies of aged macaques in the Primate Aging Study, as well as the NHP tissue bank, which has archives of frozen and fixed tissues collected from aged rhesus macaques from various colonies [7].

In addition, the primate aging database (PAD) was developed via a collaboration between NIH intramural and extramural programs [7]. This database brings together blood chemistry parameters and body weights collected from thousands of monkeys at NPRCs. The data are mostly generated from healthy animals, which can limit the utility of the findings obtained by studying these animals. The NIA has also established a Cell Bank, which contains cell lines from various NHP species as well as DNA extracted from young and aged individuals from different NHP species. These resources are available to the research community in order to promote investigations that can benefit from the use of the aged NHP model.

II. Neuroendocrine Aging in the Rhesus Macaque

Overview

Endocrine rhythms play a major role in regulating behavior and physiology functions, and their perturbation contributes to the etiology of various human pathologies. However, the underlying mechanisms for these changes are poorly understood, in large part because of the unavailability of suitable animal models for experimentation. While transgenic rodents have proven to be excellent models for systematically dissecting various mechanistic components, their clinical translational potential is often limited. Unlike humans, they are nocturnal and do not show consolidated sleep-wake patterns. Furthermore, they show significant differences in their endocrine physiology, such as with some of the release profiles of their adrenal and gonadal hormones. In contrast, rhesus macaques are relatively large animals, ideal for serial blood sampling via an implanted vascular catheter [23]. This relatively noninvasive procedure enables blood sampling to be repeatedly performed at various times across the day and night, even when the animals are asleep. This is important because many hormones show a 24-hour pattern of release, and consequently collection of blood at a single time point in the day is often insufficient to accurately assess endocrine function or to disclose age-related changes.

Adrenal Steroids

Some of the most dramatic age-related changes in these 24-hour rhythms are shown by adrenal steroids such as cortisol, dehydropiandrosterone (DHEA) and DHEA sulfate (DHEAS), which play an important role in regulating responses to stress. Elevated cortisol suppresses the immune system, breaks down tissues and has a general catabolic effect, whereas DHEA and DHEAS counter-balance the effects of cortisol by activating the immune system and building up tissues. Importantly, cortisol and DHEAS are two of the most abundant steroids in the circulation of adult humans and NHPs. Furthermore, DHEAS shows a profound age-related decrease, while cortisol levels remain constant or increase [24-28]; this is illustrated in male rhesus macaques in Figure 1 and similar age-related changes occur in females. Although the exact physiological significance of age-associated adreno-cortical hormonal changes is unclear, DHEA and DHEAS may play an important role in attenuating some of the deleterious effects of elevated cortisol levels that are associated with chronic stress. For example, in aged animals, DHEA:cortisol ratio shows a marked age-associated decline, and it has been suggested that the unopposed high cortisol levels play a key role in age-associated cognitive decline, impaired attention span, and loss of long-term memory [29, 30]. In contrast, in young adults the highly elevated DHEA and DHEAS levels help to moderate these negative effects of cortisol. In aged humans, the lower levels of DHEA and DHEAS have also been associated with cognitive disorders with a higher prevalence in the elderly, such as Alzheimer's disease [31] and depression [32]. In old men [29] and healthy postmenopausal women [33], maintaining adult-levels of endogenous DHEAS levels have been linked to better cognitive performance. While studies of the frail elderly reveal an inverse relationship between DHEAS and cognitive ability [34, 35], a comparable study in NHPs failed to disclose a similar association [36].

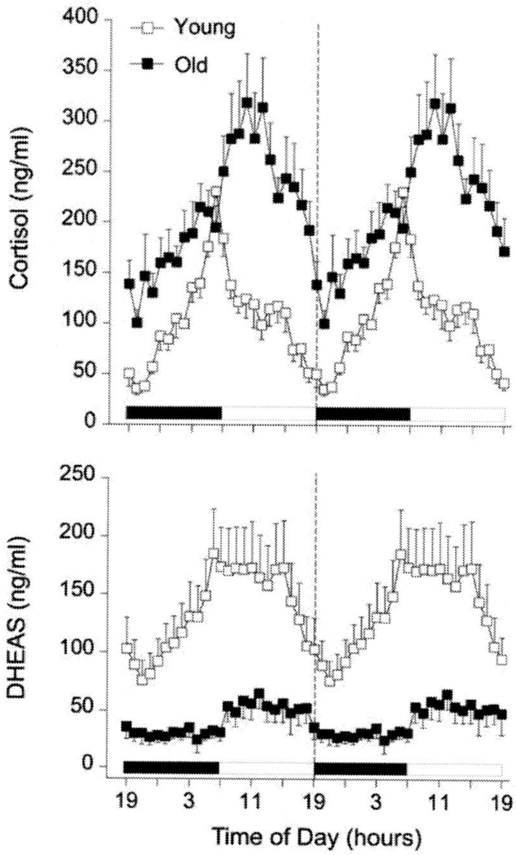

Figure. 1. Effect of age on circulating 24-hour cortisol and DHEAS profiles in male rhesus macaques. Blood samples were collected remotely every hour from 19:00 – 19:00 h, and each hormone value represents the mean ± SEM from 5 young (~10 years old) and 6 old (~26 years old) males. Note that the data have been double plotted (indicated by a vertical dashed line), to aid in the visualization of the night and day variations in hormone concentrations; the horizontal white and black bars on the abscissas correspond to the 12L:12D lighting schedule. Statistical analysis of the profiles revealed a significant age-related increase in the mean, maximum and minimum cortisol levels, but a significant decrease in these parameters for DHEAS. (From [28], with permission).

This difference could be due to the fact that frail NHP are not usually included in experimental studies. Alternatively, the authors could have missed the peak DHEA levels since they were measured at a single time point. In addition to direct actions within the central nervous system, DHEA and DHEAS may exert some of their beneficial effects indirectly, via intracrine conversion to sex steroids [37, 38].

Many organs, including the brain, appear to express the enzymes necessary for this conversion, and it is well established that sex steroids can exert neuroprotective effects in brain areas such as the hippocampus [39]. Also, it is plausible that the age-related loss of humoral circadian signaling, due to attenuated DHEA and DHEAS levels, contributes to age-related desynchronization of peripheral oscillators and exacerbation of circadian dissonance in the elderly. Although the plasma cortisol rhythm is still evident well into old age (Figure1), the elevated baseline implies that brain and peripheral organs such as the liver do not get a complete break from exposure to cortisol, which may predispose the elderly to insomnia and metabolic disorders.

Leptin

Leptin, an adipocyte-derived hormone has also been shown to have a 24-hour rhythm, with a nocturnal peak, in both lean and obese humans, as well as in individuals with type II diabetes [40, 41]. Recent studies using rhesus macaques have shown that this rhythm is circadian, because it persists even under continuous dim illumination [42]. Interestingly, the rhythm is still evident in old male macaques, whereas in peri- and post-menopausal females the difference between daytime and nighttime plasma leptin levels becomes minimal. Because leptin is generally associated with suppression of appetite it makes biological sense for its peak to occur at night, which is when humans and rhesus macaques usually sleep. Although the physiological relevance of an age-related decline is unclear, one possibility is that disruption of the circadian leptin rhythm contributes to the development of metabolic disorders and obesity [43-46]; and interferes with the maintenance of bone mass [47, 48].

Testosterone

Testosterone represents another hormone that shows a pronounced 24-hour rhythmboth in men [49-51] and in adult male rhesus macaques [52, 53]). Like leptin, the plasma testosterone rhythm shows a nocturnal peak and is associated with sleep (Figure 2); this contrasts with the timing of the daily cortisol and DHEAS peaks, which occur in the morning in association with onset of activity (Figure 1).

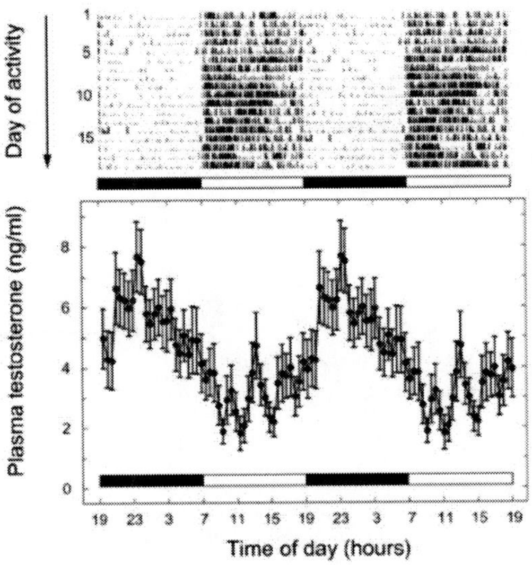

Figure. 2. Characteristic daily plasma testosterone rhythm in adult male rhesus macaques, revealed by remote serial blood sampling every 30 minutes for 24 hours. (*Upper panel*) Double-plotted actogram from a representative male rhesus macaque, obtained using an Actiwatch recorder (Philips-Respironics, Bend, OR), shows diurnal activity that is entrained to the 12L:12D light-dark cycle (indicated by the white and black horizontal bars). (*Lower panel*) Double-plotted mean plasma testosterone levels from 10 animals (± SEM). Note that in contrast to the plasma cortisol and DHEAS rhythms, which show a peak in the early morning (Figure 1), the testosterone peak occurs during the night when the animals are asleep (From Urbanski HF, Role of circadian neurendocrine rhythms in the control of behavior and physiology, Neurondrocrinology, in press, with permission).

Taken together, these endocrine data highlight the difficulty in trying to accurately assess age-associated endocrine changes based on single time point measurements. Furthermore, they emphasize the need for therapeutic hormone replacement paradigms to be carefully synchronized with underlying hormonal rhythms, so as not to perturb circadian physiological functions, such as sleep.

Ovarian Steroids

Circadian endocrine rhythms help organisms to live in an environment that shows 24-hour fluctuations in available light, ambient temperature and food availability. Moreover, rhythmic hormone output helps with temporal compartmentalization of internal biochemical processes, and like spatial compartmentalization enables many proteins to perform their cellular functions more effectively. Some endocrine rhythms, however, have a period that is much longer than 24 hours, and so it is easy to ignore the cyclic influence that these hormones may exert on physiological functions. For example the menstrual cycle in women and female rhesus macaques, is associated with monthly changes in ovarian and uterine physiology; these changes are driven by ovarian hormone synthesis, which ultimately are overtly manifested as menstrual bleeding [54, 55]. In essence, the cycle can be broken down into distinct endocrine phases (Figure 3).

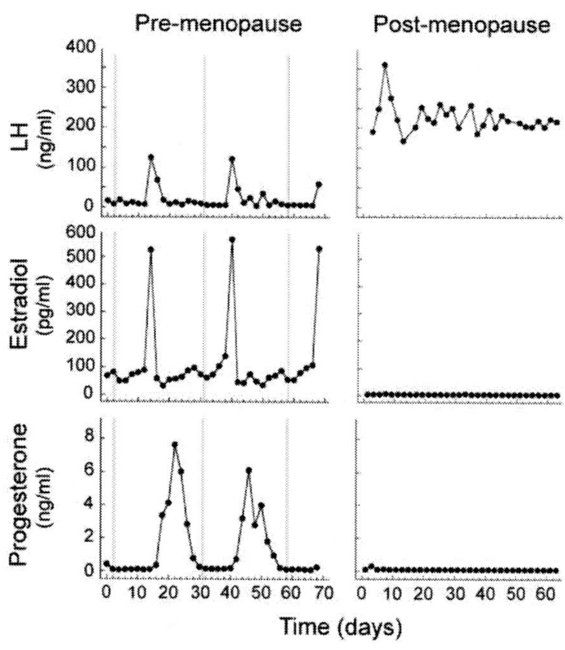

Figure. 3. Circulating reproductive hormone profiles from a representative adult pre-menopausal and a post-menopausal rhesus macaque. Blood samples were collected every 2 days for 60-70 days, and assayed for luteinizing hormone (LH), estradiol and progesterone; periodsof menstruation are represented by shaded vertical bars. Note the marked changes in sex steroid concentrations that occur across the menstrual cycle (left panels), with estradiol peaking during the late follicular phase and progesterone peaking during the mid-luteal phase. After menopause, the animals stop cycling and their sex steroid concentrations become basal; in contrast, LH concentrations become markedly elevated, due to the loss of negative sex-steroid feedback (right panels). (Data adapted from[305]).

During the follicular phase, gonadotropins -luteinizing and follicular stimulating hormones (LH and FSH respectively)- stimulate the ovarian follicles to develop and increase their secretion of estradiol. In general this takes about 2 weeks, although the length of the follicular phase tends to increase as menopause approaches and the number of estradiol-producing follicles becomes greatly diminished [55]. During the ovulatory phase, a critical threshold level of estradiol exerts a positive influence on gonadotropin releasing hormone (GnRH) release causing a surge of LH, which in turn triggers ovulation. During the luteal phase, the newly formed corpus luteum begins to secrete progesterone, which prepares the endometrium for implantation of a fertilized egg. In the absence of fertilization, however, progesterone levels decline and the endometrium lining is sloughed off. The luteal phase lasts about 2 weeks, and ends when menstrual bleeding heralds the start of another follicular phase. This cyclic chain of endocrine events, involving estradiol, gonadotropins, and progesterone represents a primary component of the reproductive axis of the female. The influence of ovarian steroids within the neuroendocrine circuits of the hypothalamus is well-established. However, these hormones have also been shown to influence other brain regions, such as the amygdala, hippocampus and prefrontal cortex, and affect serotonin, catecholaminergic, basal forebrain cholinergic neurons [39, 56-59]. Consequently, perturbation of ovarian steroid cycles, through stress, ovariectomy, or menopause, may negatively impact neuronal circuits and lead to desynchronization of physiological functions.

Up to a few years ago, it was controversial whether or not monkeys had a menopause, and the controversy was exacerbated by the lack of availability of older females. An early report in a small group of animals documented an age-related decline of the number of menstrual cycles that was accompanied by a reduced numbers of primordial and primary ovarian follicles, which placed menopause in rhesus macaques at around age 25 years [60]. A later study of neuroendocrine markers revealed a decline in ovarian steroids in old female rhesus, along with elevated gonadotropin levels, consistent with menopause [61]. However, the exact age of the females included in this study were unknown, but was estimated to be 22 years or older. Longitudinal analysis of neuroendocrine markers in female rhesus macaques that ranged from 26 to 34 years of age, also showed consistent low levels of serum estradiol and elevated LH levels that indicated the postmenopausal condition [62]. Histological examination of ovaries collected from four of these postmenopausal animals showed little evidence of folliculogenesis, indicative of exhaustion of oocytes. Another longitudinal and non-invasive study examined urine levels of estrone and progesterone metabolites in aged females of known age. This latter longitudinal study had the advantage in observing the transition into menopause and could also account for seasonal variation in reproductive cyclicity. Indeed, it was found that animals older than 25 years of age had longer inter-cycle intervals, with two animals that were post-menopausal at 29 years of age [63]. In summary, menopause does occur in the rhesus macaque, albeit late in their lifespan.

More recent studies have confirmed that old female rhesus macaques do indeed undergo menopause, characterized as a year with no menstrual cyclicity. Similar to the postmenopausal condition in women, postmenopausal rhesus macaques show a marked underlying attenuation in circulating estradiol and progesterone levels (Figure 3), which may contribute to various physiological abnormalities, including decreased bone mineral density [64-67], cognitive decline [56, 57, 68-73], and attenuated immune function [74-77]. Clinically, identification of early markers of menopause is important, because they provide an opportunity for hormonal therapies to be initiated before circulating sex steroid concentrations fall to a pathogenic level. This concern is emphasized by the results of the Women's Health Initiative, which found increased health risks in women who began hormone replacement therapy after already being postmenopausal [78-80]. Examination of the perimenopause in rhesus macaques revealed that the earliest endocrine manifestation was a significant increase in the peak plasma FSH concentrations. In addition, the concentrations of inhibin B and anti-müllerian hormone (AMH) were lower in old premenopausal animals compared to those in young adults [55]. Together, these findings represent valuable predictive knowledge about the menopausal transition period in the NHP, and lay the groundwork for postmortem studies aimed at elucidating underlying causes and immediate consequences.

In summary, the difficulty of maintaining human subjects under controlled environmental conditions, and the difficulty in performing human gene expression studies, limits progress in our understanding of how perturbed neuroendocrine rhythms contribute to physiological changes in the elderly. Therefore, the use of pragmatic animal models, such as the rhesus macaque, for mechanistic neuroendocrine studies has the potential to provide new insights into the underlying processes and to help with the development of effective therapies for age-associated human disorders.

III. Cognitive Aging in NHPs

Aging and Cognitive Decline

While rates of some leading causes of death such as heart disease, continue to decline [81], the incidence of other diseases remain somewhat constant or even show an increase (e.g., Alzheimer's Disease (AD); due in part to a lack of progress in successful treatments [82]). In addition, the data that suggests the rate of cognitive deficits increases with age is a worrisome interaction, as the number of the oldest-old continues to increase. For example, the Health and Retirement Study, a review of 1998 data showed severe cognitive deficits doubling every 5 years starting at age 70 years [83]. The AHEAD survey estimated that moderate to severe cognitive impairment was present in 9.5% of the U.S. population that was 70 years or older [84]. Indeed, the global incidence of more severe forms of neurodegenerative disease, such as AD, is growing exponentially with age, doubling every 5.5 years [85]. More recently, it was found that 22.2% of the U.S. population over the age of 70 had some form of cognitive impairment, with the exclusion of dementia [86]. Cognitive decline with age also interacts with other factors associated with poor health [87], which are predictive of higher rates of disability and mortality [88].

Clinical studies examining cognitive aging are challenged by confounds of a heterogeneous population from variable genetic backgrounds and environments. Longitudinal studies, such as the Baltimore Longitudinal Study on Aging and the Health and Retirement Study, have the advantage of examining cognitive aging trajectories over time, but are still challenged with similar difficulties. The use of animal models can control for variation in "lifestyle" (housing, level of exercise, medical care, food choice); cohort effects (different environmental conditions), selection bias (participation differentially by education level), and treatment compliance. In particular, the use of the rhesus macaque, a popular NHP model of cognitive aging has several advantages, which will be discussed below.

Age-Related Brain Changes in Rhesus Macaques

Several features of the NHP brain have been described during the past few decades, which highlight various strengths and weakness of this animal model. Unlike the rodent brain, the brain of rhesus macaques contains a high proportion of subcortical white matter and enhanced cortical folding (gyrencephalic), which is reflected in more sophisticated behaviors, [89, 90]. The more precise determination of the life history of the rhesus macaque has allowed for the description of age-related anatomical and histological changes in the brain during old age. Not surprisingly, much research was initially directed at whether or not the aged rhesus brain modeled clinical neuropathological conditions, such as AD, focusing on the classic histological biomarkers, amyloid plaques and neurofibrillary tangles.

An early light and electron microscopy study of a small group of aged rhesus brains (16-23 years old) found evidence of neuritic plaques in the cortex of the oldest of the animals [91]. Cork et al. [92] also identified amyloid deposits in the prefrontal cortex of rhesus macaques, aged 21, 27 and 30 years, some associated with neurites, but the latter appeared earlier in younger animals without associated plaques. In a paper comparing aged brains from various mammalian species, the 30-year-old rhesus brains contained amyloid plaques in the cortex as well as meningeal vessels, but without evidence of neurofibrillary tangles [93]. In a large cross-sectional study, amyloid plaques were observed in aged rhesus macaques (20-39 years old) with involvement of the cortex and blood vessels increasing with age, but again no tangles were detected [94]. The distribution of amyloid plaques was reportedly highest in the prefrontal and temporal cortices, with low levels in the occipital cortex [95]. Unlike the human condition, the shorter 40 amino acid form of amyloid is preferentially deposited in the aged rhesus brain [96]. Corroborating this accumulated evidence, was a study that found increasing amyloid burden with age in the rhesus macaque, but that there was no correlation with cognitive ability [97].

The aged rhesus brain appears to be an incomplete model of AD, showing only amyloid deposits, without the presence of neurofibrillary tangles and with no correlation of amyloid burden with cognitive ability. Moreover, a cross-sectional study of rhesus brain weight, that included almost 400 animals of both sexes, revealed that brain weight was stable into advanced old age [98]. Therefore, overt neurodegeneration is not observed by these various parameters in the aged macaque brain. Another hallmark of human brain aging was the assumption that some level of neuron loss was common (reviewed in [99]), and so it was an open question if this trend was also true in the NHP. Indeed, an early study confirmed that neuron loss with age was found in old (18-28 years old) NHPs versus young (4-7 years old) adults, which included the CA1 region of the hippocampus and prefrontal cortex [100].

During the same time period, research indicated that artifacts introduced during histological preparation resulted in differential shrinkage in younger brains, which could account for lower neuron counts density with age, and, when corrected found that there was conservation of cortical neuron number with age [101]. In addition, Terry et al., [102] utilized improved diagnostic tests for exclusion of cases with overt AD, and found that although some neuron shrank in size with age, total numbers remained stable. Thus, it seemed that for human brain aging, without the presence of severe neurodegenerative disease, morphological changes could occur, but that neuron loss was not inevitable.

The recognition of differential shrinkage of tissue and neurons with age, lead to changes in tissue processing and or the use of unbiased stereological approaches that accurately count neurons and other objects in histological preparations. Using such approaches in the aged rhesus macaque, Peters et al., [103] found neuron numbers to be stable with age in the prefrontal cortex, as well as in the striate cortex [104-106]. Another study, using stereological techniques, also verified that there was no loss of neurons in specific layers of the primary visual cortex of the rhesus macaque [107]. Examination of age-related changes in the rhesus temporal cortex also showed no neuronal loss with age in layer II [108] and layers II, III and V/VI [109] of the entorhinal cortex or in subregions of the rhesus hippocampus [110].

One study, using stereological principles also noted no neuronal loss in the aged rhesus prefrontal cortex, but a substantial 32% reduction in area 8A [111]. Thus, stability in neuron number in brain regions critical for learning and memory and the primary visual system have been described, while other cortical regions have yet to be explored.

While neuron numbers in many cortical areas appear stable with age, phenotypic changes have been reported in subcortical systems of the aged rhesus midbrain dopaminergic, serotonergic, as well as in the cholinergic neurons in the nucleus basalis. These populations, which provide critical projections to the rest of the brain appear to down-regulate with age. For example, in aged rhesus macaques, lower numbers of nigral dopaminergic neurons were counted by stereological methodology using the phenotypic markers tyrosine hydroxylase and dopamine transporter [112].

However, the level of decline, using these two markers, was significantly different. In another study, using hematoxylin and eosin stained sections, stereological counts of total cells in a cross-sectional aging study showed no change with age in the nigra of rhesus macaques [113]. However, in another study using thionin stained sections, there was an age-related decline in the number of small neurons in the nigra of old rhesus, but these were defined as GABAergic [114]. So it may be the case that down-regulation of phenotypic markers can be open to interpretation. For example, in young monkeys, manipulation of the gonadal hormone milieu has been shown to affect the expression of tyrosine hydroxylase in the nigra [115].

One age-related study of the serotoninergic system revealed that small and medium sized neurons in the nucleus centralis superior declined in old (21-23 years old) and the oldest old (20 and 33 years old) rhesus macaques [116]. This study, however, utilized Nissl-stained sections, instead of a specific serotonin marker, and depended upon location and size parameters (medium sized cells) to identify cells as being serotinergic. The decline with age of the smaller sized population was interpreted to be a loss of GABAergic neurons.

Age-related changes in the rhesus cholinergic system show interesting results, with reports of varying levels of subregional loss of neurons labeled with the phenotypic marker choline acetyl transferase in the medial septum, but hypertrophy in the size of the remaining aged neurons [117]. Cholinergic neurons in the nucleus basalis of the aged rhesus macaque, labeled by immunohistochemistry for a nerve growth factor receptor, were stable with age, but also hypertrophied in size [118].

Critically, a study in aged rhesus macaques which received nerve growth factor gene therapy, resulted in the apparent restoration of the number of cholinergic neurons to youthful levels, which suggested age-related atrophy (but not loss) of neurons occurred in this system [119]. In summary, functional atrophy of neuronal systems critical to cognitive function may be down-regulated with age, but there is the intriguing possibility that they are not lost and can be functionally restored.

Functional Anatomy of Cognition

With the situation that overt neuropathological changes do not appear to occur in the aged rhesus macaque, the aged NHP may model normative human brain aging. Declines in various cognitive domains have been described clinically, such as executive function and memory, and parallels have been sought in the monkey. However, many studies had to be conducted to see if monkeys can be behaviorally tested for equivalent cognitive domains. In addition, lesion studies, guided by clinical reports, had to be conducted to identify and then manipulate functional regions. One caveat to this approach is that the role of these brain areas was deduced by a loss of function following lesion, usually in young animals; however, it is an assumption that these studies mimic changes and potential plasticity that can occur across time as a function of aging. It should also be noted that although some tasks have been assigned functionally to specific regions of the brain, this is a simplification of the overall extent of integration.

Memory and the Temporal Lobe

The majority of the work that led to the understanding of the biological mechanisms responsible for memory is recent (reviewed in [120]). One type of memory is dependent upon the medial temporal lobe for function, and in an early study in the NHP model, lesions made in the amygdala and hippocampus resulted in severe memory problems [121]. Later, however, it was realized that the deficit more likely resulted from damage to the adjacent cortical brain areas (i.e., entorhinal, perirhinal, parahippocampal) in conjunction with hippocampal involvement [122, 123], whereas the amygdala was critical for emotional behavior [124]. The interdependence of these subregions was shown in a study where the severity of the memory impairment was increased as more subcomponents of the system were lesioned [125]. The type of memory affected by temporal lobe damage is referred to as declarative memory, which is the capacity to recollect fact and events [126]. Specific tasks to test this memory system have been developed and validated in monkeys. One such task is the delayed non-matching to sample (DNMS) test, which is a recognition memory task used in monkey models of temporal lobe lesions and aging [127]. In this task an animal is presented with a stimulus and after a delay is presented with the same stimulus and a new object. The animal must select the novel object, thus remembering the first stimulus, a task disrupted by both lesions of the medial temporal lobe and aging.

Working Memory and the Prefrontal Cortex (PFC)

The PFC in humans and NHPs has been demonstrated to be the area involved with working memory, a function that requires the ability to keep information in mind in order to complete goals by the following of rules [128]. In the NHP, the region around the principle sulcus of the dorsolateral PFC was demonstrated by a variety of lesion/functional ablation studies (reviewed in [129]) to be a key functional locus. Moreover, dopaminergic innervation is critical for normal function, and dopamine levels decline with aging [130]. Examples of tasks that target working memory are the delayed response (DR) and Wisconsin Card Sort Task (WCST), which have been adapted for use in NHPs [129]. In the former, monkeys must remember where a stimulus is located (frequently a baited well) and after a delay must correctly select the correct well from an additional well. Difficulty can be increased with longer wait times or additional wells. The WCST involves the ability to set-shift (executive function), from attention and selection of like shapes to a different criterion (e.g., select the same color, regardless of shape). Adaptation of various tasks to NHPs has been done, in order to mimic the age-related decline seen clinically.

Aging and Cognition in Macaques

There any number of excellent reviews of the effects of aging on cognitive in NHPs [127, 131, 132], including a recent overview of studies that compared the usage of different NHP species [133]. Development of tasks that are specific for tapping specific cognitive domains evolved from similar clinical tests as well as animal lesion studies that helped identify critical functional brain areas. Operationally, many of the earlier studies of cognitive aging in NHPs were conducted in the Wisconsin General Testing Apparatus (WGTA), although variations of automated systems and later computer touch-screens followed [134].

An early study, conducted in a WGTA, examined a broad age range (3-29 years old) of male and female rhesus macaques, for performance on the DNMS, a recognition task that is dependent upon the integrity of the temporal lobe [135]. In the basic task, males and females were presented with a sample object that is placed over a well that contains a food reward. After displacing the object, a wait is imposed and then the animal is presented with the prior object and a new one, with the task to displace the new object. The task was conducted with trial-unique objects, meaning that after each trial a new pair of objects was used. A "list" variation of the task presented a number of the sample objects that the animal must remember (3, 5 or 10), and then run trials with one of the sample objects presented with a novel object. In the basic task, young animals performed beyond the maximum 120 second delay, but there was a progressive age-related decline in performance as the delay time increased from 30 to 60 and 120 seconds, with the oldest animals experiencing a 12% drop in the number of correct responses versus young animals, at the longest delay. In the list task, similar results were found with an age-related decrease in performance as a function of list length and delays, with the oldest group 9% below the young group at the longest delay and longest list length. Thus, a significant effect of age on performance was seen in animals 25 years of age and older, although declines were modest and some aged animals continued to perform well.

Another study also used the trial-unique DNMS task, assessing for age differences in acquisition, delays and performance on lists in rhesus macaques [136]. Although fewer age groups were selected for this study, they were similar to an earlier study [135] in that both males and females were used and the young adult (4-5 years old) and old animals (26-27 years old) were comparable in age. The old animals were impaired in acquisition compared to young adults, at the various delays (range = 13-18%) and on the lists version of the task (range = 14-18%). However, it was noted on the latter, that some of the animals in the old group performed as efficiently as young animals. However, the performance of old animals on the lists test revealed a deficiency in recalling the last item, or recent information. Overall, these results are for the most part consistent with the idea that older animals are somewhat impaired in performing this task, but some older animals continue to function at high levels, revealing variation in cognitive skills with age.

The effect of age on DNMS was examined, comparing trial-unique objects versus repeated objects version of this task [137]. In the latter version, the same two objects were used in each trial, except the selection of the initial sample object was assigned in a pseudorandom sequence. It was found that research naive, young adult, rhesus females (9-11 years old) took significantly less training to acquire the trial-unique objects version of DNMS versus old (22-26 years old) females. However, old animals fared as well as their younger counterparts at delays that ranged from 10 sec to 22 hours. Acquisition was not different as a function of age for the repeated objects DNMS, but there was a significant decline in performance of old animals at the longer (30 and 60 sec) delays. Although prior studies [135, 136] reported significant differences on trial-unique version of DNMS, these changes were for the most part, fairly modest and the young adults used previously were younger and the old group slightly older. The author mentions that the repeated objects version of the DNMS also require a "recency" discrimination, in that the animal must remember the most recently presented object to correctly choose in the recognition phase of each trial. An age-related decline was noted in the ability of older animals to remember the latter objects on the lists variant of the trial unique DNMS [136]. To address some of these issues, a follow-up study added additional, slightly older animals (average age of 24 years) and noted that on DNMS, that only a subpopulation of older animals were impaired at longer delays (60 seconds, 2 and 10 minutes) when compared to young adults [138]. Again, the differences, although statistically different, were modest in the range of 10% and notable variation in performance was noted in the old animals.

An additional task that is sensitive to hippocampal lesions is the Delayed Recognition Span Test (DRST), which was utilized in an aging study that examined young (5-6 years old) versus old (25-27 years old) rhesus macaques [139]. Similar to DNMS, the animal must choose a non-match (novel or new) stimulus, but the task differs in that an increase in memory load is created by progressively adding an increasing array of stimuli. Performed in the WGTA, the spatial condition is composed of a 3 x 6 matrix of wells, where one well is baited with a treat and covered when the animal's view is occluded by a screen. The screen is raised and the animal dislodges the cover and retrieves the treat. Then the screen is re-lowered and the same well is covered and a new baited location is covered with an identical disk. The animal must select the new spatial location and the test keeps adding locations until an error is made. Animals were also tested under the color condition, where the same strategy is followed except a different color cover is introduced with each successful test. However, the spatial arrangement is randomized, so that the animal has to remember the previously used colors and dislodge the newest colored cover. On both versions of the DRST the aged animals performed worse than young controls and also showed a greater propensity to make perseveration errors by choosing previous locations or colors.

As mentioned above, the saliency of tests for specific species is important for the comparison of the effects of aging on cognitive function cross-species. In an interesting variation of spatial memory testing with age, Rapp et al., [140] constructed a task that approximates rodent mazes, complete with extra-maze cues. In this task young (6-8 years old) and aged (23-33 years old) rhesus macaques were trained on this open field task that consisted of an octagon with food ports at the edge of each of the eight sides. Visual cues were displayed on the walls to provide location information. As a comparison, DNMS was also tested on the same animals. While the latter task revealed no difference between age groups at various delays, it was found that in the spatial maze that the old animals made more mistakes and operated with less dependence upon the extra-maze cues. A similar study, which allowed aged rhesus macaques to freely navigate within a room, animals learned to find a treat in a series of ten food ports [141]. Initially, animals learned the task by using a serial (consecutive ports) search pattern, which evolved to a spatial strategy. After learning the task, the treat was shifted to another port and the same pattern was repeated, with an initial serial search pattern, replaced by a spatial strategy. Although this latter study had no comparison to a young group, the selection of an initial serial strategy parallels the lack of usage of extra-maze cues by older animals in the previous study [140].

While animals show a gradual decline in performance on the delay portion of the DNMS (reviewed in [132]), there is much variation between old animals, with some old animals capable of performances that rival young adults. However, in tasks that involve the PFC, there is evidence that this brain region can experience functional decline as early as middle age. One of the early studies that examined the effect of aging on short-term memory in the rhesus macaque, used an automated system that controlled for attention, motivation and allowed for self-pacing [142]. In this study, the old animals (estimated to be at least 17 years old) performed significantly worse on a DR task than young adults (3-4 years old), with a progressive decline at longer delay intervals. Importantly, parameters of the task also revealed that perceptional ability, which may decline with age, was not a factor in the testing. The same laboratory also examined the ability of reversal learning on a visual discrimination task and found that aged rhesus (estimated age 18-22 years) were able to learn new color and pattern-based discriminations, similar to young (3-6 years old) animals [143]. However, old animals showed perseveration on the reversal task, being slower to extinguish prior learning, even when normalizing for the variation in acquisition of the discrimination task.

Rapp and Amaral [137] confirmed that young (9-11 years old), experimentally naïve, female rhesus macaques performed better than older females (22-26 years old) on the DR task, at delays that ranged from 1 to 30 seconds. However, in a subsequent study that examined visual discrimination and reversal learning a year later in the same group of animals [144], the old animals performed in a similar fashion as young animals in their proficiency in reversal tasks. In addition, once old animals reached a certain level of criterion in task acquisition, the rate of learning was the same as the young adults. However, older animals took more test sessions than young animals before they achieved this more rapid phase of learning. One explanation for the enhanced performance of old animals on reversal tasks was that the animals in this study were previously tested on two versions of DNMS, which may have had a carry-over effect on the subsequent study.

Bachevalier et al., [145] in a comprehensive cross-sectional study examined four age groups (spanning 3-29 years), of male and female rhesus macaques on a spatial version of the delayed response. Using the WGTA, performance was assessed for acquisition, how effectively an animal learned the task, as well as performance across progressively longer delays. Indeed, animals had more difficulty acquiring the task as a function of increasing age, as well as in performing the task, with aged (20 years of age and older) animals performing significantly worse than the young adult group at the longer delays.

Subsequent studies of rhesus macaques have explored the role of aging on executive function. Lai et al., [146] compared spatial versus object reversal learning in young adult animals (6-11 years old) versus aged (20-28 years old), also using a WGTA, and found that both age groups acquired both tasks in a similar fashion. However, aged animals were impaired in spatial reversals, including both an initial and a second reversal. Interestingly, the same group of animals performed as well as young adults in the object reversal version of the task. Although the effect of age significantly increased perseveration errors, some aged animals performed as well as the young animals.

Executive function with age was also examined in young (5-9 years old) versus old (24-30 years old) rhesus macaques, using a computerized version of the WCST [147]. The animals were initially tested on their ability to distinguish shapes from colors and it was found that both young and aged animals could successfully perform on a three-choice discrimination task. However, older animals took more trials and made more errors in learning the initial rule, which in this case was selecting the color red when being presented with a series of pictures of three objects, while ignoring the shape of the objects. In performing the shift phase of the task, in which the rules changed successfully (select a specific shape, then a specific color and then another shape) the older animals also took more trials and made more errors. Thus, in this prefrontal cortex-based task, old animals showed increased perseveration errors. In a subsequent study, the same investigators examined male and female rhesus macaques that included a middle-aged (12-19 years old) group in addition to young (5-9 years old), or old adults (20-30 years old). Using the same task, they found that executive function started to decline by middle-age [148], hence executive function may be one of the earliest age-sensitive cognitive domains.

A recent study [149] utilized an automated cognitive test battery, the Cambridge Neuropsychological Test Automated Battery (CANTAB), which has various versions that are adapted for human and animal testing and has previously been used in rhesus macaques for the assessment of cognitive function [150-152]. This current study, however, also examined the effect of aging and compared young (average age of 7 years) and old (average age of 23 years) rhesus macaques on tests of cognitive ability. In a task that was described as being both frontal and temporal-hippocampal dependent, young animals had significant improvement over serial blocks of testing on a visuospatial task in which they had to remember the correct stimulus location out of two or three choices. Aged monkeys did not improve on this task, unless the same location/stimulus was used in a repeated fashion. Similarly, when increasing the task complexity by increasing the number of stimuli that must be remembered to two, the performance of aged animals improved with the use of repeated stimuli locations, but not as fast as young animals. Another task used in this study examined frontal lobe working memory function, the Self-Ordered Spatial Search task of the CANTAB. This task tested spatial working memory and the monkey was presented with a set of two boxes, one of which the animal must touch. The touched box briefly changes color and the screen becomes blank for a brief amount of time. The two boxes then reappear and the monkey must touch the second box to be successful. There was a significant effect of age on this task, where young animals did better than old, although both groups continued to improve with additional test blocks. Thus, similar to earlier studies, the authors conclude that aged monkeys experience a decline in performance on tasks that are reliant upon the frontal lobe.

In summary, the aged NHP exhibits progressive declines in tasks that examine memory as well as working memory performance, although the latter can appear earlier in the lifespan and the former exhibits a great deal of performance variability in the old population. This is similar to the clinical situation, where recognition memory is affected by age, but not to the extent of recollection.

However, with forms of executive function, such as working memory, aging effects are magnified with task complexity that may indicate a reduced ability to inhibit interference (see review, [153]). But, in some aged individuals, as with monkeys, performance levels can remain high [154], perhaps through increased recruitment of compensatory brain regions

Modulation of Cognitive Function with Hormone Replacement

Gonadal steroids have historically been examined for their central effects on the hypothalmo-pituitary-gonadal axis, given the presence of their cognate receptors in diencephalic structures providing feedback control. With the evolution of knowledge of steroid effects on the brain, including new receptors that function in transcriptional control, membrane effects, metabolite influence on neurotransmitter systems and extra-limbic system regulation, it is not surprising that the field has evolved to examine effects on cognition. Indeed, clinical studies have shown that in women ovariectomy decreases verbal memory, which can be ameliorated by estrogen therapy [155]. Retrospective studies also showed that hormone replacement therapy (HRT) helped reduce the risk of developing AD, but is ineffective when the disease is clinically active [156]. The recent Women's Health Initiative, in which women that were postmenopausal for more than 15 years received HRT, revealed some negative effects of delayed replacement [156]. These observations suggested that a window of opportunity for HRT, closely following menopause might be the most beneficial. In addition, different cognitive domains may not be equally affected by HRT. The focus of this section will be to review the effects of HRT on cognitive ability in aged monkeys, with many studies using similar tasks that were utilized in studies examining age-related changes in cognitive ability.

The role of gonadal hormones on cognition has a fairly recent history in studies that examined the NHP. However, the macaque model has been valuable for providing controlled experiments for various factors, such as age and the type of hormones examined. For instance, in studies examining the effects of ovariectomy on attention and cognition in *young* adult monkeys, little difference was found in ovariectomized controls versus hormone-replaced animals, which highlight the impact of age on cognitive decline [57, 157, 158]. One drawback to the macaque model is that because menopause occurs late in the NHP lifespan, surgical menopause is required in middle-aged to old animals in order to mimic the clinical situation.

Other issues arise when performing hormone replacement, due to issues of efficacy and a balance of possible negative effects along with the positive cognitive outcomes. These issues include: 1. Dosage; 2. Route of administration; 3. Duration of treatment; 4. Timing of replacement (immediate or otherwise) and 5. Type of hormones used. On the latter point, clinical studies have historically been heavily dependent Premarin (equine estrogens) and PremPro (equine estrogens and medroxyprogesterone), which are well-used pharmaceuticals but an unnatural non-primate source of steroids, which may not be optimal. NHP aging studies have utilized the naturally occurring hormones (estradiol and progesterone) in an attempt to circumvent concerns of the consequences of using unnatural, but clinically relevant (heavily prescribed) drugs.

Due in part to the late onset of menopause in monkeys, coupled with few available animals of advanced age, there has been little research performed on the relationship between menopause and cognition in the macaque. However, one report on age-matched old female rhesus macaques showed a significant impairment of performance on DR by peri-menopausal and menopausal animals versus aged animals that still had normal menstrual cycles [68]. This suggests that irregular patterns or the cessation of ovarian steroid hormones had a negative effect on PFC-based cognition. In an attempt to better define the effect of a long-term loss of ovarian hormones on cognition, Lacreuse et al., [159] compared young intact females (4-7 years old) to old intact and ovariectomized rhesus macaques (19-27 years old), with the latter group ovariectomized for an average of 12.2 years. This study showed that both the intact and long-term ovariectomized, old females were slightly impaired on DNMS, when compared to young intact controls, at 120-second delays.

At a longer delay of 600 seconds, only the long-term ovariectomized group had lower performance on this task. However, on the spatial version of the DRST, the intact old females performed statistically worse than ovariectomized old animals, which performed as well as young intact controls. However, a self-critique of this study noted that the endocrine status of the animals were not well defined, hence some of the old, intact animals may have been menopausal. In a pair of follow-up studies, the effect of oral estrogen replacement (ethinyl estradiol) was tested on the long-term ovariectomized animals, which improved performance on a hippocampal (spatial DRST), but not PFC task (DR; [160]), and had no positive effect on executive function [161]. Thus, estrogen was effective for restoring hippocampal function after many years of ovariectomy, but did not positively modulate a PFC-dependent task. This does not support the notion that hormone replacement must take place in an acute fashion, at least in the hippocampus, and that there is regional-variation in the rescue effects of estrogen.

One potential drawback of the examination of long-term ovariectomized animals, is that the timing of the loss of ovarian steroid hormones occurred when the animal was a young or middle-aged adult. More recent studies focused on surgical menopause in older rhesus females (average age of 22 years), and examined the effect of cyclic injections of estradiol cypionate on cognitive function [56, 57]. Animals were ovariectomized for an average of 30 weeks, prior to the start of treatment and were tested against young intact controls on DR and DNMS. Within 9 hour of estrogen injection, serum estradiol increased to levels typically observed at the time of the periovulatory luteinizing hormone surge, and then remained elevated at 70 pg/ml for at least the next 3 days. The effect on performance on the DR task was dramatic, with E-treated old, ovariectomized animals performing close to levels attained by young intact adults (average age of 5.2 years), whereas ovariectomized old controls performed at lower levels at all delay times [56].

On the DNMS task, the estrogen-treated old, ovariectomized animals also outperformed ovariectomized, old controls at 30- and 120-second delays. However, this was a more subtle effect as comparison to performance of young adults showed clear differences between control ovariectomized, old animals and the young intact controls, with estrogen-treated old, ovariectomized animals intermediate in performance.

As mentioned above, the performance of ovariectomized young animals on the DR task was a good as that in young ovariectomized animals receiving estrogen-treatment, overlapping with old ovariectomized animals that also received estrogen; all groups performed significantly better than old ovariectomized, vehicle-treated animals [57]. Thus, in old animals, ovariectomy had a clear negative effect on DR, which could be rescued by estrogen-treatment, at least out to 30 weeks after the loss of steroid hormones.

As discussed previously, HRT in aged animals may be more efficacious on cognition when administered soon after menopause. In addition, the type of hormone(s) used and the attained serum levels may also impact on behavioral outcomes. In an attempt to address some of these issues Voytko et al., [162, 163] examined the effect of a hormone replacement paradigm that mimicked the menstrual cycle, on cognition in old (average age of 19.7 years), acutely ovariectomized rhesus macaques. The treatment included low levels of estrogen provided by subcutaneous implants, supplemented with estradiol valerate (E) on day 12, to simulate the periovulatory surge. In an additional treatment group, estrogen was supplemented with oral progesterone (E+P) was administered for 12 days beginning on day 16, to mimic the luteal phase. Placebo controls received empty implants, and all treatments were started immediately after ovariectomy. Animals were tested on a version of DR, DMS and the WCST at 2, 12 and 24 weeks. The major findings were: 1. There was no effect of treatment on accuracy of performance on the DR task at any time point; 2. For DMS, the placebo group was less accurate than the E and E+P groups at the 30-second delay, but only at the 12 week time point; 3. On the WCST, the placebo group was impaired in shifting the cognitive set, compared to E or E+P treated animals. Therefore, immediate HRT following menopause prevented the decline of cognition, especially executive function, in the aged NHP.

To summarize, hormone replacement therapy has been examined in the old NHP and has shown promise in improving cognitive behavior, which includes varying windows of time without estrogen and different formulations of hormones and routes of administration. However, few studies have been conducted and many issues remain unresolved, especially the window of opportunity in which hormone replacement can be effective and the dose and duration of effectiveness.

Integrated with the data from aging studies, consistent effects of age and hormone loss on specific cognitive domains are beginning to emerge. While the source(s) that lead to variation in performance on cognitive aging remain to be elucidated, the NHP model is proving valuable for controlling for factors that are difficult to account for in clinical studies and by providing intriguing results that supplement clinical research. Moreover, emerging studies using this animal model are accounting for the underlying changes in physiology and functional neuroanatomy that may clearly define parameters that will prove of translational value.

IV. Aging of the Immune System

Overview

The immune system must overcome daily challenges from pathogens to protect the body from infection. The success of the immune response to infection relies on its ability to sense and evaluate microbial threats, coordinate the elimination of the threat while limiting damage to host tissues. This delicate balance is achieved through coordinated action of innate and adaptive arms of the immune system (Figure 4). The main distinguishing characteristic of these two branches of the immune response is the way they recognize antigens. Innate immunity relies on germline-encoded receptors to sense the presence of pathogens, whereas adaptive immunity utilizes a highly diverse set of receptors that are tailored to specific pathogens and generated through somatic mutations and gene recombination events. The second major defining and unique characteristic of the adaptive immune system is the development of immunological memory that manifests itself as increased functionality and frequency of responding cells upon re-exposure to the same antigen.

Aging results in several structural and functional changes in the immune system. The most evident structural changes are the involution of the thymus and the loss of bone marrow pulp. These changes lead to diminished production of lymphocytes. This age-related dysregulation in the immune system is often referred to as "immune senescence", a term first coined by Dr. Roy Walford [164-166]. Immune senescence affects both the innate and adaptive arms of the immune system, but several lines of investigation strongly suggest that defects in the adaptive arm are more severe.

One of the major consequences of immune senescence is increased susceptibility to infectious agents [167], which appear among the top ten most-common causes of death in adults in the USA aged 55 years and older [168]. This age-related increase in susceptibility to infection is especially evident when we consider newly emerging infectious diseases such as severe acute respiratory syndrome (SARS) and West Nile Virus (WNV) where most of infected patients who succumbed to infection were >50 years of age [169]. This increased susceptibility is further exacerbated by reduced vaccine efficacy in the elderly. For example, following influenza vaccine 41-58% of persons 60-74 years of age generate antibodies [170] compared to 90% in healthy adults (18-45 years old). Thus, several strategies to improve immune response to vaccination in the elderly are being explored. These include the development of new adjuvants, different vaccination schedules, and new antigens. However, it remains critical that we gain a better insight into the basic mechanisms of immune senescence in order to design new interventions aimed at improving immune function in the elderly.

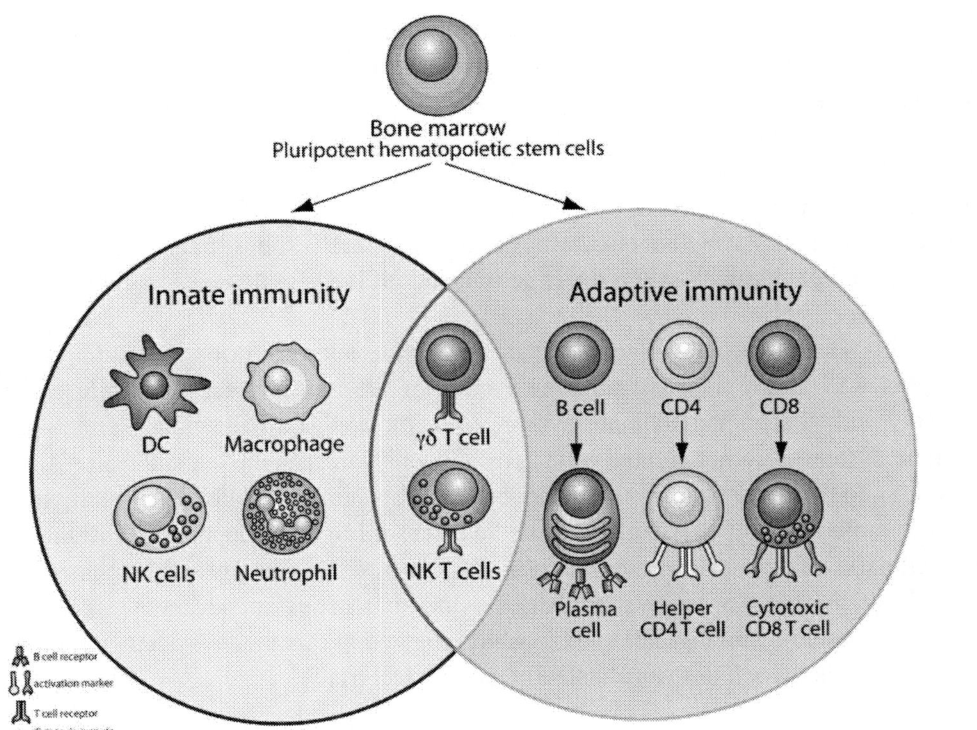

Figure 4. Overview of the immune system. The immune system can be broadly divided into innate and adaptive branches. This figure highlights the key cellular components of each branch. Innate immunity is mostly mediated by natural killer (NK), dendritic cells (DC), macrophages and neutrophils. These cells use germline-encoded receptors to recognize pathogens. Adaptive immune response is mediated by B and T cells which express antigen receptors that recognize specific pathogens. The functions of these cells are discussed in more detail throughout the review.

The vast majority of our understanding of how the immune system functions comes from studies utilizing specific pathogen free (SPF) laboratory rodents. SPF mice have been invaluable in the characterization of the cellular and molecular events that shape the development of the immune system and its response to challenges as they offer several experimental advantages, including ease of genetic manipulation and a vast array of tools and resources. Careful and methodical analyses using rodent model systems have led to key discoveries of cellular components of the immune system and have improved our understanding of how these components interact during steady state and when challenged during infection. However, there are fundamental differences between the rodent and human immune system that complicate the transfer of findings between the two species. Moreover, the inbred nature of laboratory rodents, their short lifespan and the scarcity of murine homologues of human pathogens restrict the successful transfer of immunological discoveries made in murine models to the clinical setting [171]. For instance, loss of CD28 expression on T cells especially CD8 T cells, is one of the hallmarks of immune senescence in humans. Murine T cells however, do not lose CD28 expression with age [172], thus rendering the identification of comparable T cell subsets difficult.

Aging studies using NHP have been traditionally limited to the investigation of age-related changes in behavior, cognitive function and reproduction; however, this picture is rapidly changing with the development of new immunological tools and protocols for longitudinal, systematic analysis of immunity in this animal model. Indeed, the use of NHPs to investigate immune senescence and test interventions aimed at delaying/reversing age-related changes in immune function has dramatically increased [173]. These studies have been greatly facilitated by several key advances in our understanding of the immune system of Old World monkeys, specifically the rhesus macaques (*M. mulatta*). The NHP model offers the highly desirable combination of increased life span and higher genetic homology to humans. Furthermore, markers to characterize immune cells differentiation and function are shared between NHPs and humans, which makes the NHP a robust translational model that is amenable to longitudinal extensive experimentation and simultaneous sampling of several organs.

In this section of the chapter, we will summarize our current understanding of both the innate as well as the adaptive immune system of the rhesus macaque, while emphasizing parallels with the human immune system. Our knowledge of the innate arm of the NHP immune system is rather limited, thus, we will only present the salient features of studies pertaining to this area. On the other hand, our understanding of adaptive immunity in primates is much more thorough and will therefore be discussed in more depth. We will then describe the hallmarks of immune senescence in humans and NHPs with special emphasis on studies carried out in rhesus macaques and discuss the similarities and differences. We will also describe immuno-restorative interventions investigated in this model system and discuss their efficacy and likelihood of their translation into clinical trials.

The Innate Immune System

The innate immune system is the first line of defense against pathogens and its action is mediated by several immune cell subsets that include neutrophils, natural killer (NK) cells, dendritic cells (DC), and macrophages. Because little information is available for rhesus neutrophils, they won't be discussed here. NK cells are a vital component of the innate immune system that can eliminate tumors and viral infected cells. Moreover, through the secretion of chemokines and cytokines, NK cells can modulate the kinetics and magnitude of the T cell response, and thus serve as an important link between innate and adaptive immunity. Human NK cells are identified as CD3^{neg}CD56^{pos} cells and can be subdivided based on the expression of CD16 [174].

The majority of blood and spleen resident NK cells are CD16^{pos} (85-90%); they are highly cytotoxic and secrete moderate amounts of inflammatory cytokines. The remaining NK cells are CD16^{neg}; these cells cannot kill target cells but they secrete large amounts of inflammatory cytokines [174].

Similarly, NK cells in rhesus macaques are CD3^{neg}, and can be divided into a major cytolytic population that expresses high levels of CD16 and a minor cytokine-producing population expressing low levels of CD16 (Figure 5). One major difference is that macaque NK cells, but not human NK cells, express high-levels of CD8α [175].

NK cells recognize tumor and virus-infected cells by "sensing" missing and/or altered self major histocompatibility (MHC) molecules via inhibitory and activation receptors [176]. Recognition of MHC molecules by NK cells is mediated by the killer inhibitory receptors (KIR) and CD94/NKG2 complex [177]. Homologues of the human CD94/NKG2 family members have been identified in rhesus macaques using molecular biology as well as flow cytometry tools [178].

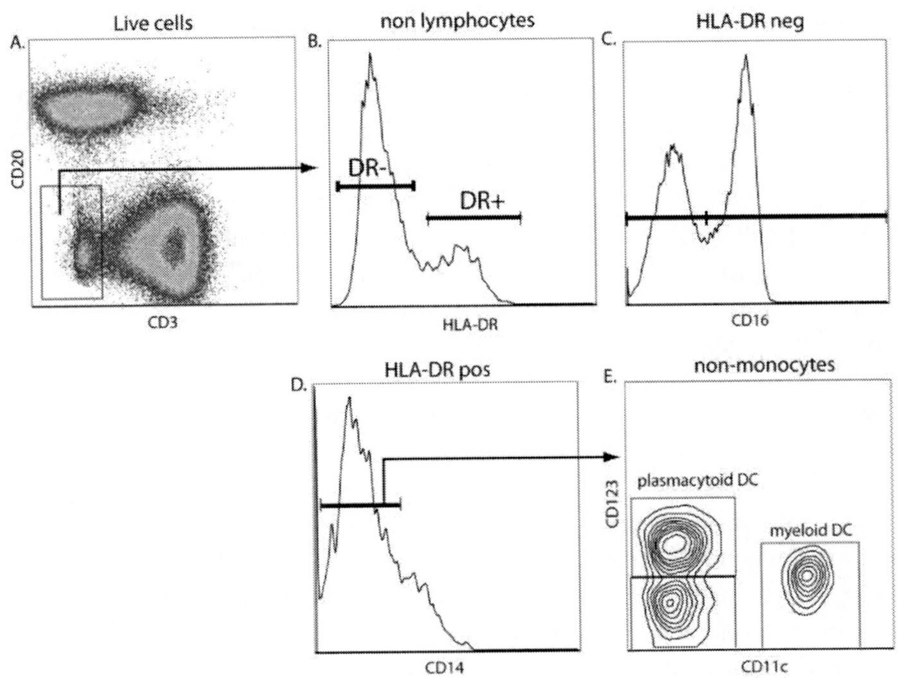

Figure 5. Identification of NK cells and DC. (A) Both NK and DC can be identified by flow cytometry by first gating on non T or B cells. (B) The non-lymphocyte population can then be subdivided based on the expression of HLA-DR molecule. (C) HLA-DR- cells contain mostly NK cells, which in turn can be divided into two subsets based on the expression of CD16. (D) The HLA-DR+ population contains monocytes/macrophages, which are CD14+ and DC, which are CD14-. (E) DC can be categorized into plasmacytoid and myeloid DC based on CD123 and CD11 molecules.

Specifically, rhesus NKp80, NKG2A, and NKG2D are expressed to similar levels as on human NK cells [179]. More importantly, cytolytic assays demonstrated that NKp80 and NKG2A displayed similar cytolytic and inhibitory functions, respectively, as their human NK counterparts [179]. Recent studies indicated that the gene expression of these markers could be modulated by simian immunodeficiency virus (SIV) infection [180]. However, the importance of NK cells function in the host response to SIV was only recently investigated thanks to the recent development of a humanized monoclonal antibody against CD16. This antibody facilitated studies to investigate the role of NK cells in the control of viral infections by making *in vivo* depletion of NK cells in rhesus macaques possible [181]. The administration of this antibody *in vivo* depleted over 88% of the NK cells (identified as $CD3^{neg}$, $CD16^{pos}$, $CD159a^{pos}$) from the circulation. Some clinical studies have suggested that NK cells can impact the progression and severity of HIV disease [182]. To test this hypothesis, rhesus macaques were infected with SIV and a subset of these animals was depleted of NK cells. NK depletion resulted in higher peak SIV viral loads especially at times when NK depletion was most notable, but did not impact viral loads during the chronic phase of the disease [181]. In macaques previously infected with SIV, administering the human anti-CD16 monoclonal antibody resulted in a shorter duration of depletion than in uninfected macaques, and it brought about little change with respect to viral loads [181]. These data suggest that NK cells do not modulate AIDS severity in rhesus macaques.

DC and macrophages play two important roles. They are scavenging cells that can ingest and destroy infectious agents in endosomal compartments. Since they capture infectious organisms with high efficiency, they also serve as professional antigen-presenting cells (APC) that are responsible for processing and presenting foreign antigen peptides to T cells thereby bridging innate and adaptive immunity. Many studies have shown that DCs are potent stimulators of T cells, therefore, a better understanding of DC physiology can lead to improved vaccine strategies. Human DCs are identified based on the absence of T, B and macrophage markers ($CD3^{neg}CD20^{neg}CD14^{neg}$) and on the presence of MHC-II molecules (a characteristic of APCs, $HLA\text{-}DR^{pos}$). DCs can then be divided into two distinct populations: 1) myeloid DCs (mDC), which are identified as $CD11c^{pos}CD123^{dim}$ [183]; and 2) plasmacytoid DCs (pDCs), which are characterized as $CD11c^{neg}CD123^{bright}$ [183]. MDCs mainly function to process and present antigens to naïve T cells and produce IL-12 upon activation; however, these cells generate very little of the antiviral cytokine interferon α (IFNα) in response to viral infection [184]. In contrast, plasmacytoid DCs, produce vast amounts of IFNα in response to viral infection [185]. Analogous DC subsets have been identified in rhesus macaques using the same surface markers [186-188] (Figure 5). More importantly, DC subsets in humans and NHPs share functional similarities and general cytokine response upon activation and viral infection [186, 189]. Since DCs represent <1% of total PBMCs in humans and NHP [186, 190], several methods have been developed to expand them *in vivo*. The administration of the hematopoietic growth factor Flt3 ligand (Flt3L) *in vivo* increases the frequency of circulating immature DCs by approximately 10-12 fold in rhesus macaques [187, 188], a finding with important implications for improved DC-targeting and anti-viral vaccine strategies [191].

The ability of APCs to recognize microbial entities relies on the detection of conserved molecular patterns referred to as pathogen-associated molecular patterns (PAMPs) by pattern recognition receptors (PRRs). The two best-characterized families of PRRs are the toll-like receptors (TLR) and the retinoic acid-inducible gene (RIG-I)-like RNA helicases (RLHs).

Following ligand recognition, these receptors initiate an intracellular signaling cascade leading to the production of several anti-microbial molecules that interfere with pathogen replication and spread. Furthermore the activation of cells of the innate immune system leads to the production of soluble mediators such as chemokines and cytokines that in turn activate the adaptive arm of the immune system [192]. Expression of TLRs within different subsets of macaque DCs exactly mimics expression in human DCs but differs significantly from murine DCs [190]. Specifically, plasmacytoid pDCs isolated from rhesus macaques express TLR-7 and TLR-9 and respond to viral infection or CpG stimulation by producing large amounts of IFNα [186, 190] and IL-12 [189, 193], and by upregulating surface expression of CD86 [189, 190, 193]. Furthermore, macaque pDCs constitutively express high levels of interferon regulatory factor-7 (IRF-7) [186], which is necessary for driving transcription of IFN-α. Similar to their human counterpart, rhesus macaque myeloid DCs, express TLR-3, -4, -7, and -8, and respond to agonists for each of the 4 TLRs by upregulating surface expression of CD40 and CD86 [190]. These studies provide important justification for the use of NHP to evaluate TLR ligands as adjuvants *in vivo*.

There has been a lot of controversy regarding how the innate immune system changes with age. Studies have suggested a reduction, maintenance, or even an increase in innate immune function with age [194]. More specifically, studies have suggested that aging is associated with increased permeability of mucosal barriers, decreased phagocytic activity of macrophages and DCs, reduced NK cell cytotoxicity, and altered production of soluble mediators such as cytokines and chemokines [194]. These alterations could lead to increased pathogen invasion and poor activation of the adaptive immune response mediated by T and B-lymphocytes. A limited number of studies have characterized age-related changes in innate immunity in NHP. For example, investigators at the Wisconsin National Primate Research Center have demonstrated that a decrease in NK cell cytolytic activity correlates strongly with a shortened lifespan in rhesus macaques, thereby serving as a useful biomarker for longevity [195]. However, changes in NK or DC subset frequency and or function remain unknown at this point.

The Adaptive Immune System

The adaptive immune branch is composed of B and T lymphocytes, which unlike cells of the innate immune system can generate responses tailored to specific pathogens. Specificity is acquired through the expression of diverse, clonally distributed antigen receptors. The initial diversity is produced in primary lymphoid organs (the thymus in the case of T cells and the bone marrow in the case of B cells) through a series of gene recombination events; further diversification occurs by somatic hypermutation of the B-cell receptor (BCR, antibody) and by functional diversification of effector T cells.

B Cell Compartment

Naïve B cells are produced in the bone marrow where B-cell precursors undergo rearrangement and expression of the heavy and light chains that constitute the BCR. Upon antigen encounter, activated naïve B cells undergo vigorous proliferation, isotype class switching (from IgM to IgG, IgA or IgE) and somatic hypermutation thereby differentiating into antigen-experienced memory B cells producing high affinity antibodies [196].

Human peripheral naïve and memory B cells can be distinguished based on the expression of CD27, a member of the tumor necrosis factor receptor family as well as the expression of IgD [196]. Using these two markers, naïve B cells are identified as CD27^{pos}IgD^{pos}, whereas memory B cells are CD27^{pos}IgD^{neg}. In addition, a third subset of B cells found in circulation expresses both CD27 and IgD and is referred to as marginal-zone like (MZ-like) B cells [197]. MZ-like B cells carry somatically mutated variable regions, which suggests that it is a transitional memory subset [198].

B cells are identified in NHP based on the expression of CD20 and MHC class II molecules (such as HLA-DR). Rhesus macaque B cells can be subdivided into three subsets as described for humans based on the expression of CD27 and IgD (Figure 6). Furthermore, activation of rhesus macaque B cells results in the up-regulation of CD80, CD86 and CD40 as described for humans [193].

A high degree of conservation of BCR heavy and light chain variable regions between humans and macaques has facilitated molecular analysis of B cell maturation in rhesus macaques [199-201]. Indeed, during simian immunodeficiency virus infection germinal centers support antigen-driven somatic hypermutation of antibodies that follows a pattern and a rate of nucleotide substitution and selection that closely mimics that described for humans infected with HIV [202].

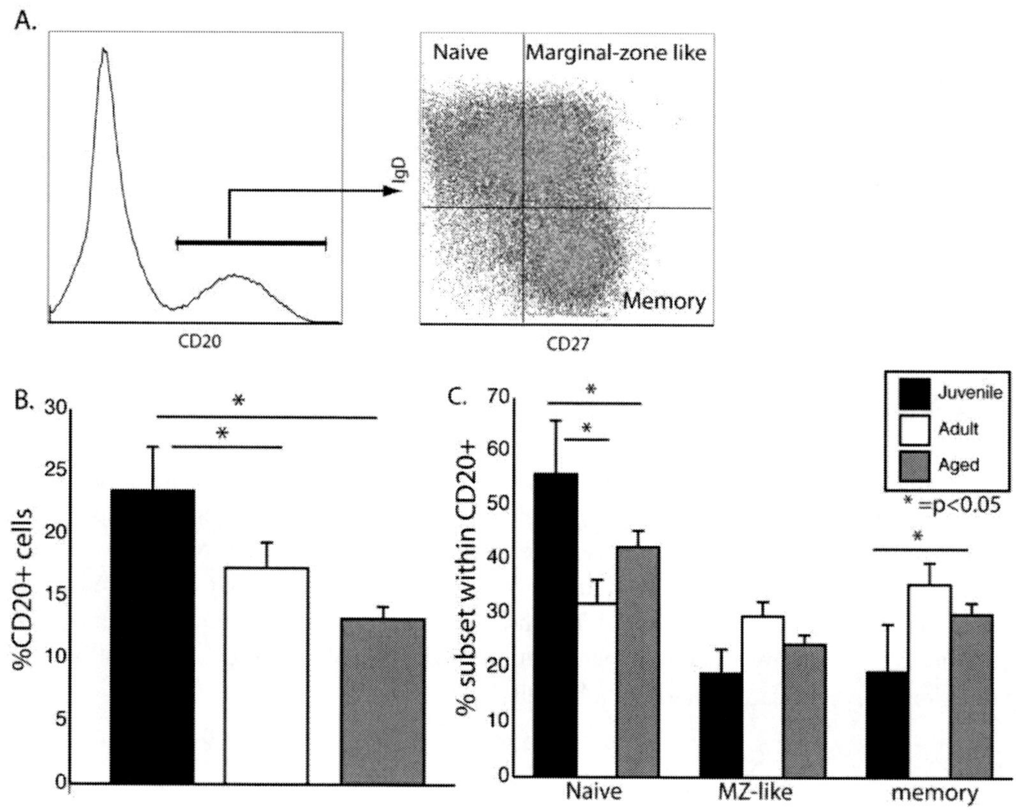

Figure 6. Age-related changes in B cell compartment. B cells are identified by flow cytometry based on the expression of CD20, and can be further subdivided into naïve, marginal-zone like and memory B cells, as described in the text. (B) Frequency of B cells decreases with age in rhesus macaques. (C) Frequency of naïve B cells decreases and that of memory subsets increases with age.

Aging is associated with a decline in B cell function and an impairment of the antibody response. Until recently, this decline was attributed to age-related reduction in CD4 T cell help to B cells. However, it has become clear that the B cell compartment undergoes several significant intrinsic changes with age that are independent from changes in the T cell compartment [203, 204]. It is widely accepted that B cell numbers decline with age [205] partly due to a decline in B cell precursor numbers. For instance, a study that examined the presence of B cell precursors in patients ranging from 2 months to 92 years of age found that despite a wide variation at all ages, statistical analysis showed a distinct decrease in B cell precursors with increasing age [206]. The data regarding B cell subsets are less clear. There is some evidence that aging results in the accumulation of memory B cells at the expense of naïve B cells [207, 208]. Regardless of changes in subset frequency, aging is accompanied by a significant decrease in the B cell repertoire diversity, which correlates strongly with frailty [209]. B cell homeostasis in NHP has not been widely investigated. Unpublished data from our lab indicate that similar to what has been observed in older humans, both male and female aged rhesus macaques (>18 years old) have fewer circulating B cells in peripheral blood than adult animals (5-10 years old) and that the frequency of antigen experienced CD27+ (memory and marginal-zone like) B cells increases with age (Figure 6).

T Cell Compartment

T cells identified based on the expression of the signaling complex CD3 are broadly divided into αβ CD4 and CD8 T cells (90%) as well as γδ T cells (10%). T cells recognize antigens in the form of small peptides bound to major histocompatibility (MHC) class I or class II molecules presented by professional APCs. CD8 T cells, commonly known as cytotoxic T cells, recognize foreign peptide bound to MHC-I molecules and have evolved to monitor for and eliminate tumor cells and those harboring intracellular pathogens (viruses and intracellular bacteria and parasites). CD4 T cells, also known as helper T cells, recognize foreign peptides bound to MHC-II and secrete a broad range of cytokines, which play a crucial role in the maturation of the B cell response as well as the CD8 T cell response.

Since the host cannot predict the precise pathogen derived peptides that it will encounter, it relies on the generation and maintenance of a diverse T cell repertoire. During T cell development, as described for BCR, T cell receptor (TCR) chains are generated through the stochastic recombination of non-contiguous TCR variable (V), diversity (D), joining (J), and constant (C) genes. T cell repertoire diversity is further enhanced by random nucleotide deletions and additions at the junctions between these segments as well as the pairing of different α and β chains. Advances in molecular analysis technology allowed us to qualitatively assess T cell repertoire diversity in rhesus macaques [210-212]. All identified genes share significant homology (~90% on average) with their human counterparts [213, 214]. In addition to structural diversity, the success of the host T cell response is modulated by functional diversity: i.e. the ability of T cells to secrete a wide array of cytokines, mediate cytolytic functions and disseminate into different organs which has been shown to be a critical determinant for disease outcome during chronic viral infections in human and NHPs [215, 216]. Significant advances in our ability to characterize functional diversity of the T cell response in rhesus macaques have been made recently. One of the most commonly used methods is intracellular cytokine staining (ICS) which allows us to measure T cell production of several cytokines as well as the expression of cytolytic molecules and tissue homing receptors simultaneously by flow cytometry [217].

The maintenance of a structurally and functionally diverse T cell repertoire is a dynamic process governed by thymic output and the exposure to antigen and cytokines, which modulate T cell survival, proliferation and death. A rigorous analysis developed over several years has demonstrated that in both humans and NHPs, CD4+ and CD8+ T cells can be broadly divided into three major subsets: naïve (Na), central memory (CM), and effector memory (EM). These subsets have been delineated in humans and NHP by the expression patterns of the cell surface markers CD28 and CD95 [218, 219] as follows: naïve ($CD28^{int}CD95^{neg}$), CM ($CD28^{pos}CD95^{pos}$) and EM ($CD28^{neg}CD95^{pos}$) (Figure 7). This differentiation has been refined based on the expression of the chemokine receptor CCR7 as described for humans with $CD95^{pos}CD28^{neg}CCR7^{neg}$ being terminally differentiated EM T cells and $CD95^{pos}CD28^{pos}CCR7^{pos}$ representing CM T cells, which leaves $CD95^{pos}CD28^{pos}CCR7^{neg}$ and $CD95^{pos}CD28^{neg}CCR7^{pos}$ as transitional populations [220] (Figure 7). In accordance with this definition, $CCR7^{pos}$ CM T cells circulate in blood and lymphoid organs but are excluded from non-lymphoid tissues such as the lung where only $CCR7^{neg}CD28^{pos}CD95^{pos}$ cells are detected [220] (Figure 7).

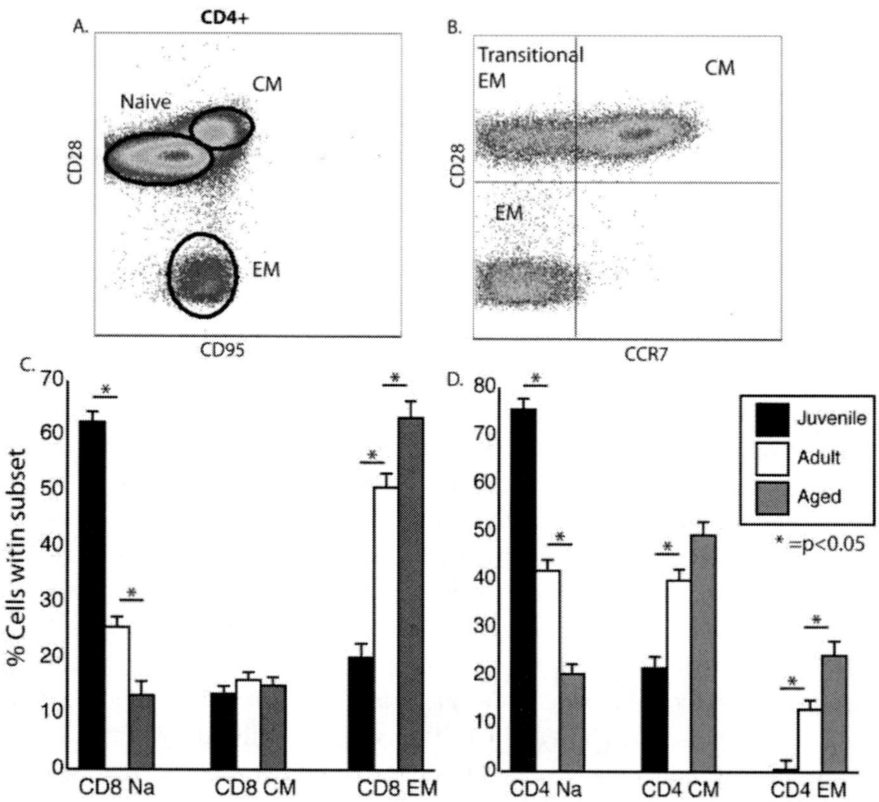

Figure 7. Age-related changes in T cell compartment. (A) CD4 and CD8 T cells can be subdivided into naïve, central (CM) and effector memory (EM), based on the expression of CD28 and CD95 markers. (B) The use of CCR7 allows the refinement of the memory T cell (CD95+ cells) subdivision and identification of transitional EM population. (C, D) Aging results in decreased naïve T cells frequency and increased EM and EM/CM populations in CD4 and CD8 T cells.

Studies have identified IL-7 and IL-15 as key players in T cell homeostasis [221]. IL-7 and MHC contact are critical for the survival of naïve T cells, where they do not induce differentiation but rather a low level of homeostatic proliferation that promotes the survival of these cells. Memory T cells on the other hand, do not require MHC contact for survival [222], but do require both IL-7 and IL-15 [223]. T cell homeostasis in rhesus macaques is governed by similar requirements as those described for rodents. The genes encoding rhesus IL-7 and IL-15 sequences share significant genetic (96%), amino acid and functional similarity with the human/rodent homologues [224]. Administration of IL-15 in rhesus macaques increased the generation of anti-viral memory T cells and their IFNγ production [225]. The administration of recombinant rhesus IL-7 induces proliferation of naïve and central memory T cells [220], whereas the administration of IL-15 enhanced proliferation of effector memory T cells [220]. These studies have provided crucial pre-clinical data for trials to improve anti-viral and anti-tumor immunity.

The most striking age-related change is the loss of naïve T cells [226] largely due to diminished thymic output, which significantly decreases after the sixth decade of life [227]. This is further exacerbated by a life-long exposure to pathogens that results in significant conversion of naïve T cells to memory T cells [228]. Clinical studies have shown that the loss of naïve T cells is accompanied by the accumulation of terminally differentiated memory T cells especially CD8 T cells that have lost CD28 expression [229, 230]. The preferential accumulation of CD28^{neg} cells within the CD8 subset may be due to the higher turnover rate of the CD8 T cells compared to CD4 T cells [229]. Several studies have shown that a significant portion of these cells are specific for the persistent herpesvirus cytomegalovirus (CMV) [231] and although some data suggest that they may play a role in controlling CMV viral burden [232], high frequencies of CD28^{neg} cells have also been associated with the poor responses to influenza vaccines [233, 234] and increased inflammation [235]. Accumulation of memory T cells results in a reduced diversity of the T cell repertoire, which is further compounded by the appearance of T cell clonal expansions (TCE) [236, 237].

Several phenomena contribute to the loss of naïve T cells. Decreased production of hematopoietic stem cells in the bone marrow leads to decreased migration of early T cell progenitors to the thymus, which in turn leads to thymic atrophy and a decline in naïve T cell production [238]. Another factor that contributes to naïve T cell loss is accelerated conversion of naïve T cells into memory T cells due to increased turnover (dubbed homeostatic proliferation). Using expression of Ki67 (a nuclear protein up-regulated with entry into cell cycle) to determine the frequency of cycling naïve and memory CD4+ T cells, homeostatic proliferation was shown to dramatically increase in individuals 70 years of age and older [227]. The increase in homeostatic proliferation in both T cell subsets was inversely related to the amount of thymic output. Thus, the progressive loss of naïve T cell production might trigger an increase in naïve T cell turnover, which increases their conversion to memory T cells and hastens loss of naïve T cells [227]. Further, a life-long exposure to chronic/persistent herpesviruses, notably CMV leads, to a continuous conversion of naïve T cells to antigen-experienced memory T cells [239].

As described for humans, rhesus macaques experience a progressive loss of naïve T cells and a concomitant accumulation of memory phenotype T cells with increasing age (Figure 7), especially CD28^{neg}CD8^{pos} EM T cells [240, 241]. Further, CD8 T cells in rhesus macaques have a propensity to undergo more rounds of divisions than CD4 T cells [218, 240]. Since extinction of CD28 expression correlates with the number of divisions, this observation could explain the selective accumulation of CD28^{neg} cells within the CD8 subset in rhesus macaques [240]. The accumulation of terminally differentiated T cells in rhesus macaques also results in narrowing of the T cell repertoire, which is compounded by the appearance of T cell clonal expansions (TCE) [242]. In fact, the frequency of naïve T cells was inversely correlated to the frequency of TCE in aged monkeys [243]. As described for humans, several factors contribute to the loss of naïve T cells in the aged rhesus macaque. In addition to thymic involution and decreased lymphopoeisis in the bone marrow, recent studies suggest that increased T cell turnover in the aged host significantly contributes to the depletion of naïve T cell reserves by increasing their conversion to memory cells [243]. A recent study compared homeostatic proliferation rate within the three major subsets of T cells of adult and aged rhesus macaques using *in vivo* BrdU pulse/chase studies. BrdU incorporation, as well as the expression of the cell cycle marker Ki-67, was elevated in peripheral naïve T cells, especially within the CD8 subset indicative on increased homeostatic proliferation. These data are consistent with clinical studies that reported an increase in CD4 T cell proliferation with age that dramatically amplified after 70 years of age [227]. Furthermore, as described for humans, we detected an increase in frequency of T cells that secrete the inflammatory cytokines IFNγ and TNFα following polyclonal stimulation in rhesus macaques [240].

The accumulation of terminally differentiated CD8 T cells leads to a decrease in the CD4:CD8 T cell ratio, which falls from ~2 to <1 in the most severe cases [244]. The enlarged CD8+ EM T cell pool is believed to be largely due to the lifelong persistent CMV infection [231]. Additional studies suggest that another persistent herpesvirus, Epstein-Barr virus (EBV), may also play a role in promoting the accumulation of terminally differentiated CD8 T cells thereby contributing to the reduced CD4:CD8 ratio [245]. One study in rhesus macaques has shown that when CD8 T cells are defined based on coincident staining with anti-CD3, the CD4:CD8 ratio in juvenile animals is ~2 [246]. This study also showed that the CD4:CD8 ratio progressively declines in rhesus macaques between 2 months and 5 years of age with no further decline detected between 5 and 7 years of age [246]. Unfortunately, no studies to date have documented changes in the CD4:CD8 ratio with increasing age. We have investigated this question by carrying out a cross-sectional analysis of the rhesus macaque colony at the Oregon National Primate Research Center (ONPRC). We measured CD4 and CD8 T cell frequencies using an anti-CD8β antibody (a marker exclusively expressed on CD8 T cells) in peripheral blood mononuclear cells (PBMC) obtained from juvenile (1-4 years old), adult (7-15 years old), and aged (18-30 years old) animals. All the animals were housed indoors and free of overt disease. As described previously [246], we also found that the CD4:CD8 ratio declines early in life and that after 7 years of age no significant changes are observed (Messaoudi et. al., unpublished observations). However, the relationship between persistent herpesviruses and the CD4:CD8 ratio in aged rhesus macaques is unclear at this point. Rhesus CMV (RhCMV) is ubiquitous in most colonies where the majority of animals seroconvert by 1 year of age [247, 248]. Similar to HCMV infection in humans, RhCMV infection results in the development of large T cell clonal responses in rhesus macaques [249, 250]. However, careful analysis of the CMV status (antibody titer, frequency of antigen specific T cells) and how it correlates with the CD4:CD8 ratio has not yet been carried out in rhesus macaques.

The impact of aging on cytokine production by T cells is controversial with some studies reporting a decrease in inflammatory cytokine production while others describing an increase or no change [251, 252]. Despite these differences, there are trends that suggest that T cell inflammatory cytokine production increases with age in humans. Early studies showed that T cells from aged individuals secrete higher levels of the pro-inflammatory cytokines IL-6, TNF-α and IFN-γ following long stimulations (24-72 hrs) with phorbol 12-myristate 13-acetate (PMA) [253-255]. Later studies using CD8+ T cells purified from PBMC obtained from healthy young and old individuals also showed an age-related increase in the number of central and effector memory CD8+ T cells producing type-1 (IFNγ, IL-2, TNFα) and type-2 cytokines (IL-4, IL-6, IL-10) cytokines in response to PMA stimulation [256]. On the other hand, studies investigating CD4+ T cell cytokine production in response to PMA showed a decrease in the frequency of IFNγ, and TNFα-producing cells and an increase in IL-4-producing cells [257]. Thus, in the case of CD4 T cells, aging seems to result in a shift from type-1 to type-2 cytokine production profile [257]. T cell cytokine profiles in the elderly were also influenced by CMV status [258]. T cells from CMV^{neg} donors showed increased production of IL-2, IL-4 and IFNγ with age, whereas T cells from CMV^{pos} individuals only showed an increased in IFNγ production with age compared to CMV^{neg} young and middle-aged individuals [258]. Thus chronic CMV infection not only modulates T cell homeostasis, but also T cell cytokine production. It is possible that the lack of increase of IL-2 and IL-4 production in CMV^{pos} donors in the face of increased IFNγ production could contribute to inflammatory processes in the elderly.

In contrast to the vast literature describing cytokine production by human PBMC/T cells, very few studies have examined the impact of age on T cell cytokine production in NHP. LPS or PHA stimulation of whole blood or purified PBMC from rhesus macaques showed an age-related increase in IL-10 and IL-6, a decrease in IFNγ production and no change in IL-1β or TNFα production by cytokine ELISA [259, 260]. However, in these studies, cytokine production was determined by ELISA in tissue culture supernatant and therefore the exact contribution of T cells to cytokine production is unknown. More recent studies specifically investigated the impact of age on T cell production of IFNγ and TNFα by splenocytes from neonate (>1 year old), juvenile (1-3 year old), adult (5-10 year old), and aged animals (>18 years old) using intracellular cytokine staining (ICS) following anti-CD3 stimulation [240]. Data from these studies showed an age-related increase in the frequency of IFNγ- and TNFα-secreting T cells primarily amongst the CD8+ EM T cell population. Unpublished studies from our laboratory have shown the same phenomenon using PBMC instead of splenocytes. However, more experiments need to be carried out to investigate age-related changes in type 2 cytokines such as IL-4, IL-6 and IL-10 in NHPs and the impact of CMV status on age-related changes in cytokine production in rhesus macaques.

In addition to altered cytokine production, immune senescence results in decreased T cell proliferative capacity as measured by the ability of the T cells to respond to polyclonal stimulation [261]. This decreased proliferative potential is believed to be due in part to the accumulation of terminally differentiated $CD28^{neg}$ T cells, which have severely shortened telomeres [262] as well as a reduced ability to induce telomerase [263], resulting in an impaired ability to proliferate [264]. Similar studies in rhesus macaques have shown that the ability of T cells to proliferate *in vitro* in response to polyclonal stimulation, such as anti-CD3, is reduced in aged rhesus macaques (>18 years old) compared to adult monkeys (5-10 years old) [240]. Not only was the percentage of cells that enter cell cycle reduced but the number of divisions was also reduced. Thus, as described for humans, aging leads to decreased T cell proliferation capacity in rhesus macaques.

Response to Vaccination/Infection

Changes associated with immune senescence are believed to contribute to increased morbidity and mortality from infectious diseases. As early as 1987 studies by Ershler W.B. and colleagues showed that lymphocyte proliferation and antibody production to tetanus toxoid vaccination were reduced in older monkeys, especially males. In fact, lymphocytes from old male monkeys responded significantly less to test mitogens than did those of old female or young males and females [265]. Another study also showed a compromised mucosal immune response following cholera toxin vaccination [266]. Similarly, we have recently compared the immune response of young and aged female rhesus macaques to modified vaccinia ankara (MVA) and found that aged macaques experience a decline in CD8 T cell and antibody responses to MVA vaccine [77, 267]. The decrease in CD8 T cell responses in aged macaques correlated with the severity of naïve T cell loss [267]. In contrast to these findings, another recent study found that aged (19-24 years old) baboons generated a more robust antibody response to a Yersinia antigen than young baboons [268]. One possible explanation for the difference in outcome is that the study by Stacy and colleagues use 2.5 years old animals as young controls while previous studies used adult animals (10-year-old macaques). Juvenile animals could potentially have an immature immune system that can result in reduced immune responses to vaccination and/or infection compared to adult animals.

Rejuvenation of the Aged Immune System

Caloric Restriction

One intervention that showed promise in revitalizing the immune system, and as such, has been the focus of recent studies is caloric restriction (CR) [269]. In 1915 and 1917, Osborne and Mendel observed that decreasing food intake by female rats delayed the onset of fertility and extended their lifespan [270]. These observations were then confirmed in 1935 by McCay [270]. Since then, several studies refined these findings and demonstrated that a caloric reduction of 30% was sufficient to achieve the same life extension benefits. Today, most CR diets contain supplementation with vitamins and essential elements to prevent nutritional deficiency and ensure that the intervention results in caloric restriction and not malnutrition [271]. CR delays the aging process in several short-lived species such as rodents, nematodes and yeast.

Furthermore, CR delayed the onset/reduced the severity of immune senescence in rodents. Specifically, CR preserved frequency of naïve T cells [272], maintained T cell proliferative capacity [273], most likely by enhancing apoptosis of senescent cells [274] and also improved the T cell response to influenza [275]. CR reduced the levels of circulating IL-6 and TNFα [276, 277], which in turn decreased the incidence of autoimmune diseases [278, 279] and cancer [280, 281]. Several epidemiological studies strongly suggested that CR would have similar benefits in humans [282, 283]. To rigorously address this question, studies in NHPs were initiated at the NIA [284] as well as the Wisconsin National Primate Research Center [285]. Reports from both studies showed that CR exerted many of the beneficial physiological effects observed in rodents such as improved cardiovascular and glucoregulatory function [286, 287]. Interestingly, other studies have shown that in contrast to the beneficial effects observed in rhesus macaques, CR did not result in decreased weight, oxidative damage or glycooxidation in squirrel monkeys. These observations suggest that CR is not protective in New World monkeys [14].

We have recently investigated the impact of CR on T cell senescence in a rhesus macaque cohort established at the NIA [242, 288]. Our studies examined the severity of T cell senescence in male and female rhesus macaques that have been on CR since early age (1-2 years), early adulthood (5-7 years) or advanced age (>17 years) [242, 288]. We showed that adult-onset CR preserved naïve T cells, a diverse T cell repertoire, T cell proliferative capacity and reduced the frequency of memory T cells that secreted pro-inflammatory factors IFNγ and TNFα in response to CD3 stimulation [242]. These findings strongly suggest that immune response to infection/vaccination would be more robust in CR animals.

However, these findings differed from earlier reports that suggested a decrease in immune function in CR rhesus macaques [289-291]. The duration of CR as well as the age at onset differed significantly between the earlier studies and our reported data. We examined animals that started caloric restriction during early adulthood and were maintained continuously on this diet for 14 years, whereas earlier studies examined animals after a short period of CR (~2 years) and whose age at onset was somewhere between 8 and 14 years old.

To determine whether age at CR onset modulates its impact on immune senescence, we measured several immune senescence parameters in animals who were calorically restricted either early (1-2 years) or late (>17 years) in life [288]. Our analysis revealed that juvenile-onset CR (JO-CR) in male rhesus macaques resulted in a significant increase in the frequency of terminally differentiated effector memory CD4 and CD8 T cells and the reduction of T cell repertoire diversity. Furthermore JO-CR increased the frequency of CD4 and CD8 T cells that secreted IFNγ and TNFα in response to CD3 stimulation, and reduced T cell proliferative capacity. Late onset CR resulted in reduced T cell proliferative response to stimulation in the absence of any detectable changes in T cell subset distribution. Taken together these data strongly suggest that there is an optimal window for the initiation of CR. The delay of immune senescence by adult onset CR would theoretically lead to improvement, whereas juvenile onset CR would be expected to result in a diminished immune response to infection or vaccination. Studies are currently underway to address this critical question.

IL-7 Therapy

The loss of naïve T cells is probably the most dramatic change that occurs with increasing age. Since naïve T cells are the host's reserve to respond to new pathogens, there is increased focus on thymic regeneration in the elderly [292]. One of the leading candidates considered is IL-7 therapy. As discussed earlier, IL-7 promotes the survival of thymocytes, naïve and central memory T cells. Studies in rodents showed that IL-7 administration enhanced reconstitution of peripheral T cell repertoire [293, 294], and improved thymic output in aged mice [295]. More recent studies tested whether IL-7 treatment can overcome CD4 depletion in SIV infected NHP [296]. These findings prompted studies in aged rhesus macaques to determine whether IL-7 treatment could improve thymic output and the immune response to vaccination [297]. The administration of recombinant IL-7 to aged rhesus macaques resulted in a transient increase in the number of circulating CD4 and CD8 T cells that returned to baseline 10 weeks after the last treatment dose. IL-7 administration also increased thymic output as measured by the frequency of T cell receptor excision circles (TRECs). These episomal circular DNA fragments are the result of gene recombination events that take place at the TCR locus. Since TRECs are not amplified during cell division, they are often used as an indirect measure of thymic output [298]. Interestingly, despite a transient increase in TREC numbers, there was no change in the frequency of naïve T cells following IL-7 treatment [297]. This phenomenon can be explained by rapid conversion of recent thymic emigrants to central memory T cells [299].

Alternatively, the administration of IL-7 could promote the trafficking of recent thymic emigrants and naïve T cells from lymphoid tissue into the blood. IL-7 treated animals also generated a more robust antibody response following vaccination with influenza [297]. However, the antibody titer eventually decreased to the same set point as that observed in control animals. These preliminary studies suggest a promising role for IL-7 treatment but the optimal schedule of administration and dosage remain to be determined. A recent clinical study showed that IL-7 administration to refractory cancer patients resulted in the same outcomes reported for the aged primates including increased number of circulating CD4 and CD8 T cells, higher TREC numbers and improved T cell repertoire diversity [300].

Keratinocyte Growth Factor

Another intervention aimed at rejuvenating the thymus is the administration of human keratinocyte growth factor (KGF). Initial studies in rodents showed that the KGF protein has considerable potential to increase thymopoeisis and improve T-cell dependent antibody responses in aged mice [301]. This outcome prompted recent studies in primates to evaluate the ability of KGF to promote thymic regeneration following irradiation and stem cell transplantation. The results from the study show that KGF-treated animals showed a well-preserved thymic structure and output which translated into a broader T cell repertoire and improved antibody responses compared to control animals [302, 303].

Thymosin

Thymosin alpha 1 (Tα1) has been investigated for more than 25 years as a possible solution for the decreased immune function observed in both the elderly and the immune suppressed. In early NHP studies female rhesus macaques (18-25 years old) were treated with Tα1 or placebo after being vaccinated against tetanus toxoid [265]. Although an increase in lymphocyte proliferation and NK cells cytotoxic activity was observed in Tα1 treated animals, no significant effect on antibody response to the tetanus vaccine was observed [265]. However, encouraged by the increased lymphocyte proliferation, Tα1 was recently tested as an adjuvant to influenza vaccination in individuals aged 65 years and older [304]. Amongst those patients receiving Tα1 following influenza vaccination, 69% (31/45) had a fourfold increase in influenza antibody titers compared to 52% (21/40) of the placebo group. In a second trial by the same group involving 330 elderly volunteers the authors observed only a modest reduction in the number of influenza cases in the presence of Tα1, but those who developed influenza showed only mild or nonexistent symptoms compared to the placebo group [304], suggesting that Tα1 treatment might attenuate disease severity.

In summary, the decline of immune function with age is believed to be responsible for increased morbidity and mortality. Unfortunately, we do not yet have a test that signals the onset of immune senescence other than age. Advances in our understanding of the aging process in rhesus immune system coupled with improved methodologies to manipulate different cellular components of this model will certainly accelerate our ability to uncover mechanisms of and develop novel vaccine strategies and therapeutics that are better suited for vulnerable populations such as the elderly.

Conclusion

The US Census Bureau projects that from 2030 onwards, the proportion of age 65 and older will be 20%. More importantly, it also projects that the population aged 85 and over will grow to nearly 21 million by 2050 (http://factfinder.census.gov). Therefore, the identification and validation of biomarkers of successful aging in an out-bred animal model are essential for our ability to predict the efficacy of interventions aimed at promoting health aging. We have highlighted advances in three major areas of aging research made possible with use of aged NHPs, with special emphasis on the rhesus macaque, a popular model for biomedical research. Importantly, characterization and experimentation using this model continues in many research areas, thus further strengthening its value. However, because of limited availability of aged NHPs, future studies will involve increased multidisciplinary approaches, which will undoubtedly lead to a better understanding of the interaction of general physiological changes with specific aging domains. In addition, we look forward to the identification and validation of biomarkers of aging that can help predict functional outcomes for ameliorative and preventative treatments. Furthermore, an explanation for the variation in functional outcome as a factor of aging is of interest, as it suggests that the maintenance of performance can be manipulated, either as a "replacement" paradigm or perhaps by the regulation of compensatory mechanisms. Understanding the causes of and prevention of fragility in the aged NHP model will lay the foundation for the development of creative solutions for aging-associated disorders in humans thereby extending our "healthspan".

References

[1] US Census Bureau, Population Division, 2010. http://www.census.gov/cgi-bin/broker
[2] Trewavas A. Malthus foiled again and again. *Nature* 2002 Aug 8;418(6898):668-70.
[3] US Census Bureau, Population Estimates, 2010. http://www.census.gov/popest/national/asrh/NC-EST2009-sa.html
[4] Thompson WS. Population. *Am. J. Sociol.* 1929;34:959-75.
[5] Caldwell JC. Demographers and the study of mortality: scope, perspectives, and theory. *Ann. N Y Acad. Sci.* 2001 Dec;954:19-34.

[6] Wagner JE, Kavanagh K, Ward GM, Auerbach BJ, Harwood HJ, Jr., Kaplan JR. Old world nonhuman primate models of type 2 diabetes mellitus. *ILAR J.* 2006;47(3):259-71.

[7] Nadon NL. Of mice and monkeys: National Institute on Aging resources supporting the use of animal models in biogerontology research. *J. Gerontol. A. Biol. Sci. Med. Sci.* 2006;61(8):813-5.

[8] Register TC. Primate models in women's health: inflammation and atherogenesis in female cynomolgus macaques (Macaca fascicularis). *Am. J.Primatol.* 2009;71(9):766-75.

[9] Shively CA, Clarkson TB. The unique value of primate models in translational research. *Am. J. Primatol.* 2009;71(9):715-21.

[10] Kennedy RC, Shearer MH, Hildebrand W. Nonhuman primate models to evaluate vaccine safety and immunogenicity. *Vaccine* 1997;15(8):903-8.

[11] Gray WL. Simian varicella: a model for human varicella-zoster virus infections. *Rev. Med. Virol.* 2004;14(6):363-81.

[12] Ward KW. Recent advances in pharmacokinetic extrapolation from preclinical data to humans. *Expert Opin. Drug Metab. Toxicol.* 2005;1(4):583-94.

[13] Gardner MB, Luciw PA. Macaque models of human infectious disease. *ILAR J.* 2008;49(2):220-55.

[14] Williams L. Aging Cebidae. *Interdiscip Top Gerontol* 2008;36:49-61.

[15] Sequencing RMG, Consortium A, Gibbs RA, Rogers J, Katze MG, Bumgarner R, et al. Evolutionary and biomedical insights from the rhesus macaque genome. *Science* ~~2007~~ April 13, 2007;316(5822):222-34.

[16] Rawlins RG, Kessler MJ, Turnquist JE. Reproductive performance, population dynamics and anthropometrics of the free-ranging Cayo Santiago rhesus macaques. *J. Med. Primatol.* 1984;13(5):247-59.

[17] Kessler MJ, Berard JD, Rawlins RG, Bercovitch FB, Gerald MS, Laudenslager ML, et al. Tetanus antibody titers and duration of immunity to clinical tetanus infections in free-ranging rhesus monkeys (Macaca mulatta). *Am. J. Primatol.* 2006;68(7):725-31.

[18] Tigges J, Gordon TP, McClure HM, Hall EC, Peters A. Survival rate and life span of rhesus monkeys at the Yerkes Regional Primate Research Center. *Am. J. Primatol.* 1988;15(3):263-73.

[19] Dyke B, Gage TB, Mamelka PM, Goy RW, Stone WH. A demographic analysis of the Wisconsin Regional Primate Center rhesus colony, 1962–1982. *Am. J. Primatol.,* 1986;10(3):257-69.

[20] Bodkin NL, Alexander TM, Ortmeyer HK, Johnson E, Hansen BC. Mortality and morbidity in laboratory-maintained rhesus monkeys and effects of long-term dietary restriction. *J. Gerontol. A. Biol. Sci. Med. Sci.* 2003;58(3):212-9.

[21] Colman RJ, Anderson RM, Johnson SC, Kastman EK, Kosmatka KJ, Beasley TM, et al. Caloric restriction delays disease onset and mortality in rhesus monkeys. *Science* 2009;325(5937):201-4.

[22] VandeBerg JL, Williams-Blangero S. Advantages and limitations of nonhuman primates as animal models in genetic research on complex diseases. *J. Med. Primatol.* 1997;26(3):113-9.

[23] Urbanski HF. Circadian variation in the physiology and behavior of humans and nonhuman primates. In: Raber J, editor. *Animal Models of Behavioral Analysis*. NY: Springer, 2011: 217-35.
[24] Wise PM. Neuroendocrine correlates of aging. In: Conn PM, Freeman ME, editors. *Neuroendocrinology in Physiology and Medicine*. Totowa: Human Press Inc., 1999: 371-87.
[25] Orentreich N, Brind JL, Vogelman JH, Andres R, Baldwin H. Long-term longitudinal measurements of plasma dehydroepiandrosterone sulfate in normal men. *J. Clin. Endocrinol. Metab.* 1992;75(4):1002-4.
[26] Labrie F, Belanger A, Cusan L, Gomez JL, Candas B. Marked decline in serum concentrations of adrenal C19 sex steroid precursors and conjugated androgen metabolites during aging. *J. Clin. Endocrinol. Metab.* 1997;82(8):2396-402.
[27] Urbanski HF, Downs JL, Garyfallou VT, Mattison JA, Lane MA, Roth GS, et al. Effect of caloric restriction on the 24-hour plasma DHEAS and cortisol profiles of young and old male rhesus macaques. *Ann. N Y Acad. Sci.* 2004;1019:443-7.
[28] Downs JL, Mattison JA, Ingram DK, Urbanski HF. Effect of age and caloric restriction on circadian adrenal steroid rhythms in rhesus macaques. *Neurobiol. Aging,* 2008;29(9):1412-22.
[29] van Niekerk JK, Huppert FA, Herbert J. Salivary cortisol and DHEA: association with measures of cognition and well-being in normal older men, and effects of three months of DHEA supplementation. *Psychoneuroendocrinology* 2001;26(6):591-612.
[30] Karishma KK, Herbert J. Dehydroepiandrosterone (DHEA) stimulates neurogenesis in the hippocampus of the rat, promotes survival of newly formed neurons and prevents corticosterone-induced suppression. *Eur. J. Neurosci.* 2002;16(3):445-53.
[31] Weill-Engerer S, David JP, Sazdovitch V, Liere P, Eychenne B, Pianos A, et al. Neurosteroid quantification in human brain regions: comparison between Alzheimer's and nondemented patients. *J. Clin. Endocrinol. Metab.* 2002;87(11):5138-43.
[32] Michael A, Jenaway A, Paykel ES, Herbert J. Altered salivary dehydroepiandrosterone levels in major depression in adults. *Biol. Psychiatry,* 2000 Nov 15;48(10):989-95.
[33] Davis SR, Shah SM, McKenzie DP, Kulkarni J, Davison SL, Bell RJ. Dehydroepiandrosterone sulfate levels are associated with more favorable cognitive function in women. *J. Clin. Endocrinol. Metab.* 2008;93(3):801-8.
[34] Morrison MF, Katz IR, Parmelee P, Boyce AA, TenHave T. Dehydroepiandrosterone sulfate (DHEA-S) and psychiatric and laboratory measures of frailty in a residential care population. *Am. J. Geriatr. Psychiatry,* 1998;6(4):277-84.
[35] Morrison MF, Redei E, TenHave T, Parmelee P, Boyce AA, Sinha PS, et al. Dehydroepiandrosterone sulfate and psychiatric measures in a frail, elderly residential care population. *Biol. Psychiatry,* 2000 15;47(2):144-50.
[36] Herndon JG, Lacreuse A, Ladinsky E, Killiany RJ, Rosene DL, Moss MB. Age-related decline in DHEAS is not related to cognitive impairment in aged monkeys. *Neuroreport* 1999;10(17):3507-11.
[37] Labrie F. Intracrinology. *Mol. Cell Endocrinol.* 1991;78(3):C113-8.
[38] Sorwell KG, Urbanski HF. Dehydroepiandrosterone and age-related cognitive decline. *Age* (Dordr) 2010;32(1):61-7.
[39] McEwen B. Estrogen actions throughout the brain. *Recent Prog. Horm. Res.* 2002;57:357-84.

[40] Sinha MK, Ohannesian JP, Heiman ML, Kriauciunas A, Stephens TW, Magosin S, et al. Nocturnal rise of leptin in lean, obese, and non-insulin-dependent diabetes mellitus subjects. *J. Clin. Invest.* 1996;97(5):1344-7.

[41] Simon C, Gronfier C, Schlienger JL, Brandenberger G. Circadian and ultradian variations of leptin in normal man under continuous enteral nutrition: relationship to sleep and body temperature. *J. Clin. Endocrinol. Metab.* 1998;83(6):1893-9.

[42] Downs JL, Urbanski HF. Aging-related sex-dependent loss of the circulating leptin 24-h rhythm in the rhesus monkey. *J. Endocrinol.* 2006 Jul;190(1):117-27.

[43] Laughlin GA, Yen SS. Hypoleptinemia in women athletes: absence of a diurnal rhythm with amenorrhea. *J. Clin. Endocrinol. Metab.* 1997;82(1):318-21.

[44] Matkovic V, Ilich JZ, Badenhop NE, Skugor M, Clairmont A, Klisovic D, et al. Gain in body fat is inversely related to the nocturnal rise in serum leptin level in young females. *J. Clin. Endocrinol. Metab.* 1997;82(5):1368-72.

[45] Balligand JL, Brichard SM, Brichard V, Desager JP, Lambert M. Hypoleptinemia in patients with anorexia nervosa: loss of circadian rhythm and unresponsiveness to short-term refeeding. *Eur. J. Endocrinol.* 1998;138(4):415-20.

[46] Stoving RK, Vinten J, Handberg A, Ebbesen EN, Hangaard J, Hansen-Nord M, et al. Diurnal variation of the serum leptin concentration in patients with anorexia nervosa. *Clin. Endocrinol.* (Oxf) 1998;48(6):761-8.

[47] Ducy P, Amling M, Takeda S, Priemel M, Schilling AF, Beil FT, et al. Leptin inhibits bone formation through a hypothalamic relay: a central control of bone mass. *Cell,* 2000;100(2):197-207.

[48] Elefteriou F, Ahn JD, Takeda S, Starbuck M, Yang X, Liu X, et al. Leptin regulation of bone resorption by the sympathetic nervous system and CART. *Nature* 2005;434(7032):514-20.

[49] Bremner WJ, Vitiello MV, Prinz PN. Loss of circadian rhythmicity in blood testosterone levels with aging in normal men. *J. Clin. Endocrinol. Metab.* 1983;56(6):1278-81.

[50] Tenover JS, Matsumoto AM, Clifton DK, Bremner WJ. Age-related alterations in the circadian rhythms of pulsatile luteinizing hormone and testosterone secretion in healthy men. *J. Gerontol.* 1988;43(6):M163-9.

[51] Cooke RR, McIntosh JE, McIntosh RP. Circadian variation in serum free and non-SHBG-bound testosterone in normal men: measurements, and simulation using a mass action model. *Clin. Endocrinol.* (Oxf) 1993;39(2):163-71.

[52] Schlatt S, Pohl CR, Ehmcke J, Ramaswamy S. Age-related changes in diurnal rhythms and levels of gonadotropins, testosterone, and inhibin B in male rhesus monkeys (Macaca mulatta). *Biol. Reprod.* 2008;79(1):93-9.

[53] Sitzmann BD, Leone EH, Mattison JA, Ingram DK, Roth GS, Urbanski HF, et al. Effects of moderate calorie restriction on testosterone production and semen characteristics in young rhesus macaques (Macaca mulatta). *Biol. Reprod.* 2010;83(4):635-40.

[54] Zeleznik AJ, Pohl CR. Control of follicular development, copus luteum function, the maternal recognition of pregnancy and the neuroendorine regulation of the menstrual cycle in higher primates. In: Neil JD, editor. *Physiology of Reproduction.* 3rd ed. St. Louis: Elsevier, 2006: 2449-510.

[55] Downs JL, Urbanski HF. Neuroendocrine changes in the aging reproductive axis of female rhesus macaques (Macaca mulatta). *Biol. Reprod.* 2006;75(4):539-46.
[56] Rapp PR, Morrison JH, Roberts JA. Cyclic estrogen replacement improves cognitive function in aged ovariectomized rhesus monkeys. *J. Neurosci.* 2003;23(13):5708-14.
[57] Hao J, Rapp PR, Janssen WG, Lou W, Lasley BL, Hof PR, et al. Interactive effects of age and estrogen on cognition and pyramidal neurons in monkey prefrontal cortex. *Proc. Natl Acad Sci U S A* 2007;104(27):11465-70.
[58] Noriega NC, Eghlidi DH, Garyfallou VT, Kohama SG, Kryger SG, Urbanski HF. Influence of 17beta-estradiol and progesterone on GABAergic gene expression in the arcuate nucleus, amygdala and hippocampus of the rhesus macaque. *Brain Res.*;1307:28-42.
[59] Eghlidi DH, Haley GE, Noriega NC, Kohama SG, Urbanski HF. Influence of age and 17beta-estradiol on kisspeptin, neurokinin B, and prodynorphin gene expression in the arcuate-median eminence of female rhesus macaques. *Endocrinology*;151(8):3783-94.
[60] Catchpole HR, Van Wagenen G. Reproduction in the rhesus monkey, *Macaca mulatta*. In: Bourne GH, editor. *The Rhesus Monkey*. N.Y: Academic Press, 1975: 117-40.
[61] Hodgen GD, Goodman AL, O'Connor A, Johnson DK. Menopause in rhesus monkeys: model for study of disorders in the human climacteric. *Am. J. Obstet. Gynecol.* 1977 15;127(6):581-4.
[62] Walker ML. Menopause in female rhesus monkeys. *Am. J. Primatol.* 1995;35(1):59-71.
[63] Gilardi KV, Shideler SE, Valverde CR, Roberts JA, Lasley BL. Characterization of the onset of menopause in the rhesus macaque. *Biol. Reprod.* 1997;57(2):335-40.
[64] Black A, Tilmont EM, Handy AM, Scott WW, Shapses SA, Ingram DK, et al. A nonhuman primate model of age-related bone loss: a longitudinal study in male and premenopausal female rhesus monkeys. *Bone* 2001;28(3):295-302.
[65] Colman RJ, Kemnitz JW, Lane MA, Abbott DH, Binkley N. Skeletal effects of aging and menopausal status in female rhesus macaques. *J. Clin. Endocrinol. Metab.* 1999;84(11):4144-8.
[66] Seifert-Klauss V, Mueller JE, Luppa P, Probst R, Wilker J, Hoss C, et al. Bone metabolism during the perimenopausal transition: a prospective study. *Maturitas* 2002;41(1):23-33.
[67] Ebeling PR, Atley LM, Guthrie JR, Burger HG, Dennerstein L, Hopper JL, et al. Bone turnover markers and bone density across the menopausal transition. *J. Clin. Endocrinol. Metab.* 1996;81(9):3366-71.
[68] Roberts JA, Gilardi KV, Lasley B, Rapp PR. Reproductive senescence predicts cognitive decline in aged female monkeys. *Neuroreport,* 1997 May 27;8(8):2047-51.
[69] Lacreuse A. Effects of ovarian hormones on cognitive function in nonhuman primates. *Neuroscience* 2006;138(3):859-67.
[70] Halbreich U, Lumley LA, Palter S, Manning C, Gengo F, Joe SH. Possible acceleration of age effects on cognition following menopause. *J. Psychiatr. Res.* 1995;29(3):153-63.
[71] Paganini-Hill A, Henderson VW. Estrogen deficiency and risk of Alzheimer's disease in women. *Am. J. Epidemiol.* 1994;140(3):256-61.
[72] Shaywitz SE, Shaywitz BA, Pugh KR, Fulbright RK, Skudlarski P, Mencl WE, et al. Effect of estrogen on brain activation patterns in postmenopausal women during working memory tasks. *JAMA* 1999;281(13):1197-202.

[73] Tinkler GP, Tobin JR, Voytko ML. Effects of two years of estrogen loss or replacement on nucleus basalis cholinergic neurons and cholinergic fibers to the dorsolateral prefrontal and inferior parietal cortex of monkeys. *J. Comp. Neurol.* 2004;469(4):507-21.

[74] Porter VR, Greendale GA, Schocken M, Zhu X, Effros RB. Immune effects of hormone replacement therapy in post-menopausal women. *Exp. Gerontol.* 2001;36(2):311-26.

[75] Keller ET, Zhang J, Yao Z, Qi Y. The impact of chronic estrogen deprivation on immunologic parameters in the ovariectomized rhesus monkey (Macaca mulatta) model of menopause. *J. Reprod. Immunol.* 2001;50(1):41-55.

[76] Pfeilschifter J, Koditz R, Pfohl M, Schatz H. Changes in proinflammatory cytokine activity after menopause. *Endocr. Rev.* 2002;23(1):90-119.

[77] Engelmann F, Barron A, Urbanski H, Neuringer M, Kohama SG, Park B, et al. Accelerated immune senescence and reduced response to vaccination in ovariectomized female rhesus macaques. *Age* (Dordr) 2011.

[78] Grodstein F, Manson JE, Stampfer MJ. Hormone therapy and coronary heart disease: the role of time since menopause and age at hormone initiation. *J. Womens Health* (Larchmt), 2006;15(1):35-44.

[79] Klaiber EL, Vogel W, Rako S. A critique of the Women's Health Initiative hormone therapy study. *Fertil Steril.* 2005;84(6):1589-601.

[80] Prentice RL, Langer R, Stefanick ML, Howard BV, Pettinger M, Anderson G, et al. Combined postmenopausal hormone therapy and cardiovascular disease: toward resolving the discrepancy between observational studies and the Women's Health Initiative clinical trial. *Am. J. Epidemiol.* 2005;162(5):404-14.

[81] Xu J, Kochanek KD, Murphy SL, Tejada-Vera B. Deaths: final data for 2007. *National Vital Statistics Reports*, 2010: 58: 1-136.

[82] Statistics Federal Interagency Forum on Aging-Related Statistics. *Older Americans 2010: Key indicators of well-being*. 2010.

[83] Aging NIA. Growing Older in America: The Health and Retirement. 2007.

[84] Suthers K, Kim JK, Crimmins E. Life expectancy with cognitive impairment in the older population of the United States. *J. Gerontol. B. Psychol. Sci. Soc. Sci.* 2003;58(3):S179-86.

[85] Ziegler-Graham K, Brookmeyer R, Johnson E, Arrighi HM. Worldwide variation in the doubling time of Alzheimer's disease incidence rates. *Alzheimers Dement* 2008;4(5):316-23.

[86] Plassman BL, Langa KM, Fisher GG, Heeringa SG, Weir DR, Ofstedal MB, et al. Prevalence of cognitive impairment without dementia in the United States. *Ann. Intern. Med. 2008*;148(6):427-34.

[87] Stewart ST, Zelinski EM, Wallace RB. Age, medical conditions, and gender as interactive predictors of cognitive performance: the effects of selective survival. *J. Gerontol. B. Psychol. Sci. Soc. Sci.* 2000;55(6):P381-3.

[88] McGuire LC, Ford ES, Ajani UA. Cognitive functioning as a predictor of functional disability in later life. *Am. J. Geriatr. Psychiatry,* 2006;14(1):36-42.

[89] Fukuda S, del Zoppo GJ. Models of focal cerebral ischemia in the nonhuman primate. *ILAR J.* 2003;44(2):96-104.

[90] Rogers J, Kochunov P, Zilles K, Shelledy W, Lancaster J, Thompson P, et al. On the genetic architecture of cortical folding and brain volume in primates. *Neuroimage,* 2010;53(3):1103-8.

[91] Wisniewski HM, Ghetti B, Terry RD. Neuritic (senile) plaques and filamentous changes in aged rhesus monkeys. *J. Neuropathol. Exp. Neurol.* 1973;32(4):566-84.

[92] Cork LC, Masters C, Beyreuther K, Price DL. Development of senile plaques. Relationships of neuronal abnormalities and amyloid deposits. *Am. J. Pathol.* 1990;137(6):1383-92.

[93] Selkoe DJ, Bell DS, Podlisny MB, Price DL, Cork LC. Conservation of brain amyloid proteins in aged mammals and humans with Alzheimer's disease. *Science* 1987;235(4791):873-7.

[94] Uno H, Alsum PB, Dong S, Richardson R, Zimbric ML, Thieme CS, et al. Cerebral amyloid angiopathy and plaques, and visceral amyloidosis in aged macaques. *Neurobiol. Aging,* 1996;17(2):275-81.

[95] Struble RG, Price DL, Jr., Cork LC, Price DL. Senile plaques in cortex of aged normal monkeys. *Brain Res.* 1985;361(1-2):267-75.

[96] Gearing M, Tigges J, Mori H, Mirra SS. A beta40 is a major form of beta-amyloid in nonhuman primates. *Neurobiol. Aging,* 1996;17(6):903-8.

[97] Sloane JA, Pietropaolo MF, Rosene DL, Moss MB, Peters A, Kemper T, et al. Lack of correlation between plaque burden and cognition in the aged monkey. *Acta Neuropathol.* 1997;94(5):471-8.

[98] Herndon JG, Tigges J, Klumpp SA, Anderson DC. Brain weight does not decrease with age in adult rhesus monkeys. *Neurobiol. Aging,* 1998;19(3):267-72.

[99] Wickelgren I. For the cortex, neuron loss may be less than thought. *Science* 1996;273(5271):48-50.

[100] Brizzee KR, Ordy JM, Bartus RT. Localization of cellular changes within multimodal sensory regions in aged monkey brain: possible implications for age-related cognitive loss. *Neurobiol. Aging,* 1980 Summer;1(1):45-52.

[101] Haug H, Kuhl S, Mecke E, Sass NL, Wasner K. The significance of morphometric procedures in the investigation of age changes in cytoarchitectonic structures of human brain. *J. Hirnforsch.* 1984;25(4):353-74.

[102] Terry RD, DeTeresa R, Hansen LA. Neocortical cell counts in normal human adult aging. *Ann. Neurol.* 1987;21(6):530-9.

[103] Peters A, Leahu D, Moss MB, McNally KJ. The effects of aging on area 46 of the frontal cortex of the rhesus monkey. *Cereb. Cortex,* 1994;4(6):621-35.

[104] Vincent SL, Peters A, Tigges J. Effects of aging on the neurons within area 17 of rhesus monkey cerebral cortex. *Anat. Rec.* 1989;223(3):329-41.

[105] Peters A, Sethares C. Aging and the Meynert cells in rhesus monkey primary visual cortex. *Anat. Rec.* 1993;236(4):721-9.

[106] Peters A, Nigro NJ, McNally KJ. A further evaluation of the effect of age on striate cortex of the rhesus monkey. *Neurobiol. Aging,* 1997;18(1):29-36.

[107] Hof PR, Nimchinsky EA, Young WG, Morrison JH. Numbers of meynert and layer IVB cells in area V1: a stereologic analysis in young and aged macaque monkeys. *J. Comp. Neurol.* 2000;420(1):113-26.

[108] Gazzaley AH, Thakker MM, Hof PR, Morrison JH. Preserved number of entorhinal cortex layer II neurons in aged macaque monkeys. *Neurobiol. Aging,* 1997;18(5):549-53.

[109] Merrill DA, Roberts JA, Tuszynski MH. Conservation of neuron number and size in entorhinal cortex layers II, III, and V/VI of aged primates. *J. Comp. Neurol.* 2000;422(3):396-401.

[110] Keuker JI, Luiten PG, Fuchs E. Preservation of hippocampal neuron numbers in aged rhesus monkeys. *Neurobiol. Aging,* 2003;24(1):157-65.

[111] Smith DE, Rapp PR, McKay HM, Roberts JA, Tuszynski MH. Memory impairment in aged primates is associated with focal death of cortical neurons and atrophy of subcortical neurons. *J. Neurosci.* 2004;24(18):4373-81.

[112] Emborg ME, Ma SY, Mufson EJ, Levey AI, Taylor MD, Brown WD, et al. Age-related declines in nigral neuronal function correlate with motor impairments in rhesus monkeys. *J. Comp. Neurol.* 1998;401(2):253-65.

[113] Pakkenberg H, Andersen BB, Burns RS, Pakkenberg B. A stereological study of substantia nigra in young and old rhesus monkeys. *Brain Res.* 1995;693(1-2):201-6.

[114] Siddiqi Z, Kemper TL, Killiany R. Age-related neuronal loss from the substantia nigra-pars compacta and ventral tegmental area of the rhesus monkey. *J. Neuropathol. Exp. Neurol.* 1999;58(9):959-71.

[115] Leranth C, Roth RH, Elsworth JD, Naftolin F, Horvath TL, Redmond DE, Jr. Estrogen is essential for maintaining nigrostriatal dopamine neurons in primates: implications for Parkinson's disease and memory. *J. Neurosci.* 2000;20(23):8604-9.

[116] Kemper TL, Moss MB, Rosene DL, Killiany RJ. Age-related neuronal loss in the nucleus centralis superior of the rhesus monkey. *Acta Neuropathol.* 1997;94(2):124-30.

[117] Stroessner-Johnson HM, Rapp PR, Amaral DG. Cholinergic cell loss and hypertrophy in the medial septal nucleus of the behaviorally characterized aged rhesus monkey. *J. Neurosci.* 1992;12(5):1936-44.

[118] Voytko ML, Sukhov RR, Walker LC, Breckler SJ, Price DL, Koliatsos VE. Neuronal number and size are preserved in the nucleus basalis of aged rhesus monkeys. *Dementia* 1995;6(3):131-41.

[119] Smith DE, Roberts J, Gage FH, Tuszynski MH. Age-associated neuronal atrophy occurs in the primate brain and is reversible by growth factor gene therapy. *Proc. Natl. Acad. Sci. U S A,* 1999;96(19):10893-8.

[120] Squire LR. Memory and brain systems: 1969-2009. *J. Neurosci.* 2009;29(41):12711-6.

[121] Mishkin M. Memory in monkeys severely impaired by combined but not by separate removal of amygdala and hippocampus. *Nature* 1978 May 25;273(5660):297-8.

[122] Murray EA, Mishkin M. Object recognition and location memory in monkeys with excitotoxic lesions of the amygdala and hippocampus. *J. Neurosci.* 1998;18(16):6568-82.

[123] Squire LR, Zola-Morgan S. The medial temporal lobe memory system. *Science* 1991;253(5026):1380-6.

[124] Zola-Morgan S, Squire LR, Alvarez-Royo P, Clower RP. Independence of memory functions and emotional behavior: separate contributions of the hippocampal formation and the amygdala. *Hippocampus* 1991;1(2):207-20.

[125] Zola-Morgan S, Squire LR, Ramus SJ. Severity of memory impairment in monkeys as a function of locus and extent of damage within the medial temporal lobe memory system. *Hippocampus* 1994;4(4):483-95.

[126] Squire LR, Clark RE, Bayley PJ. Medial temporal lobe function and memory. In: Gazzaniga MS, editor. *Cognitive Neuroscience III*, 2004: 691-708.

[127] Gallagher M, Rapp PR. The use of animal models to study the effects of aging on cognition. *Annu. Rev. Psychol.* 1997;48:339-70.

[128] Miller EK, Cohen JD. An integrative theory of prefrontal cortex function. *Annu. Rev. Neurosci,* 2001;24:167-202.

[129] Castner SA, Goldman-Rakic PS, Williams GV. Animal models of working memory: insights for targeting cognitive dysfunction in schizophrenia. *Psychopharmacology* (Berl) 2004;174(1):111-25.

[130] Goldman-Rakic PS. Prenatal formation of cortical input and development of cytoarchitectonic compartments in the neostriatum of the rhesus monkey. *J. Neurosci.* 1981;1(7):721-35.

[131] Albert MS, Moss MB. Neuropsychology of aging: findings in humans and monkeys. In: Schneider EL, editor. *Handbook of the Biology of Aging*, 1996: 217-33.

[132] Peters A, Rosene DL, Moss MB, Kemper TL, Abraham CR, Tigges J, et al. Neurobiological bases of age-related cognitive decline in the rhesus monkey. *J. Neuropathol. Exp. Neurol.* 1996;55(8):861-74.

[133] Lacreuse A, Herndon JG. Nonhuman primate models of cognitive aging. In: Bizon JL, Woods A, editors. *Animal Models of Cognitive Aging*: Humana Press, 2009: 29-58.

[134] Mandell DJ, Sackett GP. Comparability of developmental cognitive assessments between standard and computer testing methods. *Dev. Psychobiol.* 2009;51(1):1-13.

[135] Presty SK, Bachevalier J, Walker LC, Struble RG, Price DL, Mishkin M, et al. Age differences in recognition memory of the rhesus monkey (Macaca mulatta). *Neurobiol. Aging*, 1987;8(5):435-40.

[136] Moss MB, Rosene DL, Peters A. Effects of aging on visual recognition memory in the rhesus monkey. *Neurobiol. Aging,* 1988;9(5-6):495-502.

[137] Rapp PR, Amaral DG. Evidence for task-dependent memory dysfunction in the aged monkey. *J. Neurosci.* 1989;9(10):3568-76.

[138] Rapp PR, Amaral DG. Recognition memory deficits in a subpopulation of aged monkeys resemble the effects of medial temporal lobe damage. *Neurobiol. Aging,* 1991;12(5):481-6.

[139] Moss MB, Killiany RJ, Lai ZC, Rosene DL, Herndon JG. Recognition memory span in rhesus monkeys of advanced age. *Neurobiol. Aging,* 1997;18(1):13-9.

[140] Rapp PR, Kansky MT, Roberts JA. Impaired spatial information processing in aged monkeys with preserved recognition memory. *Neuroreport,* 1997;8(8):1923-8.

[141] Haley GE, Landauer N, Renner L, Weiss A, Hooper K, Urbanski HF, et al. Circadian activity associated with spatial learning and memory in aging rhesus monkeys. *Exp. Neurol.* 2009;217(1):55-62.

[142] Bartus RT, Fleming D, Johnson HR. Aging in the rhesus monkey: debilitating effects on short-term memory. *J. Gerontol.* 1978;33(6):858-71.

[143] Bartus RT, Dean RL, 3rd, Fleming DL. Aging in the rhesus monkey: effects on visual discrimination learning and reversal learning. *J. Gerontol.* 1979;34(2):209-19.

[144] Rapp PR. Visual discrimination and reversal learning in the aged monkey (Macaca mulatta). *Behav. Neurosci.* 1990;104(6):876-84.

[145] Bachevalier J, Landis LS, Walker LC, Brickson M, Mishkin M, Price DL, et al. Aged monkeys exhibit behavioral deficits indicative of widespread cerebral dysfunction. *Neurobiol. Aging.* 1991;12(2):99-111.

[146] Lai ZC, Moss MB, Killiany RJ, Rosene DL, Herndon JG. Executive system dysfunction in the aged monkey: spatial and object reversal learning. *Neurobiol. Aging,* 1995;16(6):947-54.

[147] Moore TL, Killiany RJ, Herndon JG, Rosene DL, Moss MB. Impairment in abstraction and set shifting in aged rhesus monkeys. *Neurobiol. Aging,* 2003;24(1):125-34.

[148] Moore TL, Killiany RJ, Herndon JG, Rosene DL, Moss MB. Executive system dysfunction occurs as early as middle-age in the rhesus monkey. *Neurobiol. Aging,* 2006;27(10):1484-93.

[149] Nagahara AH, Bernot T, Tuszynski MH. Age-related cognitive deficits in rhesus monkeys mirror human deficits on an automated test battery. *Neurobiol. Aging.* 2010;31(6):1020-31.

[150] Taffe MA, Weed MR, Gold LH. Scopolamine alters rhesus monkey performance on a novel neuropsychological test battery. *Brain Res. Cogn. Brain Res.* 1999;8(3):203-12.

[151] Weed MR, Taffe MA, Polis I, Roberts AC, Robbins TW, Koob GF, et al. Performance norms for a rhesus monkey neuropsychological testing battery: acquisition and long-term performance. *Brain Res. Cogn. Brain Res.* 1999;8(3):185-201.

[152] Weed MR, Bryant R, Perry S. Cognitive development in macaques: attentional set-shifting in juvenile and adult rhesus monkeys. *Neuroscience*, 2008;157(1):22-8.

[153] Drag LL, Bieliauskas LA. Contemporary review 2009: cognitive aging. *J. Geriatr. Psychiatry Neurol.*;23(2):75-93.

[154] Moss MB, Moore TL, Schlettler SP, Killiany R, Rosene DL. Successful vs. unsuccessful aging in the rhesus monkey. In: Riddle DR, editor. *Brain Aging, Models, Methods and Mechanisms.* Boca Raton: CRC press, 2007.

[155] Sherwin BB, Henry JF. Brain aging modulates the neuroprotective effects of estrogen on selective aspects of cognition in women: a critical review. *Front. Neuroendocrinol.* 2008;29(1):88-113.

[156] Henderson VW. Cognitive changes after menopause: influence of estrogen. *Clin. Obstet. Gynecol.* 2008;51(3):618-26.

[157] Voytko ML. The effects of long-term ovariectomy and estrogen replacement therapy on learning and memory in monkeys (Macaca fascicularis). *Behav. Neurosci.* 2000;114(6):1078-87.

[158] Lacreuse A, Wilson ME, Herndon JG. No effect of different estrogen receptor ligands on cognition in adult female monkeys. *Physiol. Behav.* 2009;96(3):448-56.

[159] Lacreuse A, Herndon JG, Moss MB. Cognitive function in aged ovariectomized female rhesus monkeys. *Behav. Neurosci.* 2000;114(3):506-13.

[160] Lacreuse A, Wilson ME, Herndon JG. Estradiol, but not raloxifene, improves aspects of spatial working memory in aged ovariectomized rhesus monkeys. *Neurobiol. Aging.* 2002;23(4):589-600.

[161] Lacreuse A, Chhabra RK, Hall MJ, Herndon JG. Executive function is less sensitive to estradiol than spatial memory: performance on an analog of the card sorting test in ovariectomized aged rhesus monkeys. *Behav. Processes.* 2004;67(2):313-9.

[162] Voytko ML, Higgs CJ, Murray R. Differential effects on visual and spatial recognition memory of a novel hormone therapy regimen of estrogen alone or combined with progesterone in older surgically menopausal monkeys. *Neuroscience* 2008;154 (4):1205-17.

[163] Voytko ML, Murray R, Higgs CJ. Executive function and attention are preserved in older surgically menopausal monkeys receiving estrogen or estrogen plus progesterone. *J. Neurosci.* 2009;29(33):10362-70.

[164] Effros RB. Roy Walford and the immunologic theory of aging. *Immun. Ageing.* 2005 Apr 25;2(1):7.

[165] McElhaney JE, Effros RB. Immunosenescence: what does it mean to health outcomes in older adults? *Curr. Opin. Immunol.* 2009;21(4):418-24.

[166] Walford RL. *The immunologic theory of aging*. Copenhagen: Munksgaard, 1969.

[167] Weinberger B, Herndler-Brandstetter D, Schwanninger A, Weiskopf D, Grubeck-Loebenstein B. Vaccines: biology of immune responses to vaccines in elderly persons. *Clin. Infect.s Dis.* 2008;46(7):1078-84.

[168] High KP. Infection as a cause of age-related morbidity and mortality. *Ageing Res. Rev.* 2004;3(1):1-14.

[169] Chen Q, Liang WN, Liu GF, Liu M, Xie XQ, Wu J, et al. Case fatality rate of severe acute respiratory syndromes in Beijing. *Biomed. Environ. Sci.* 2005 Aug;18(4):220-6.

[170] Goodwin K, Viboud C, Simonsen L. Antibody response to influenza vaccination in the elderly: a quantitative review. *Vaccine* 2006;24(8):1159-69.

[171] Hein WR, Griebel PJ. A road less travelled: large animal models in immunological research. *Nat. Rev. Immunol.* 2003;3(1):79-84.

[172] Clambey ET, van Dyk LF, Kappler JW, Marrack P. Non-malignant clonal expansions of CD8+ memory T cells in aged individuals. *Immunol. Rev.* 2005;205(1):170-89.

[173] Nikolich-Zugich J. Non-human primate models of T-cell reconstitution. *Seminars in Immunology* 2007;19(5):310-7.

[174] Lanier LL. Up on the tightrope: natural killer cell activation and inhibition. *Nat. Immunol.* 2008;9(5):495-502.

[175] Ramothea L. Webster RPJ. Delineation of multiple subpopulations of natural killer cells in rhesus macaques. *Immunology* 2005;115(2):206-14.

[176] Christopher E. Andoniou JDC, Mariapia A. Degli-Esposti,. Killers and beyond: NK-cell-mediated control of immune responses. *Eur. J. Immunol.* 2008;38(11):2938-42.

[177] Biassoni R. Natural killer cell receptors. *Adv. Exp. Med. Biol.* 2008;640:35-52.

[178] LaBonte ML, Letvin NL. Variable NKG2 expression in the peripheral blood lymphocytes of rhesus monkeys. *Clin. Exp. Immunol.* 2004;138(2):205-12.

[179] Mavilio D, Benjamin J, Kim D, Lombardo G, Daucher M, Kinter A, et al. Identification of NKG2A and NKp80 as specific natural killer cell markers in rhesus and pigtailed monkeys. *Blood* 2005 Sep 1;106(5):1718-25.

[180] LaBonte ML, McKay PF, Letvin NL. Evidence of NK cell dysfunction in SIV-infected rhesus monkeys: Impairment of cytokine secretion and NKG2C/C2 expression. *Eur. J. Immunol.* 2006;36(9):2424-33.

[181] Choi EI, Reimann KA, Letvin NL. In vivo natural killer cell depletion during primary simian immunodeficiency virus infection in rhesus monkeys. *J. Virol.* , 2008;82(13):6758-61.

[182] Alter G, Altfeld M. NK cells in HIV-1 infection: evidence for their role in the control of HIV-1 infection. *J. Intern. Med.* 2009;265(1):29-42.
[183] Olweus J, BitMansour A, Warnke R, Thompson PA, Carballido J, Picker LJ, et al. Dendritic cell ontogeny: a human dendritic cell lineage of myeloid origin. *Proc. Natl. Acad. Sci. U S A,* 1997;94(23):12551-6.
[184] Siegal FP, Kadowaki N, Shodell M, Fitzgerald-Bocarsly PA, Shah K, Ho S, et al. The nature of the principal type 1 interferon-producing cells in human blood. *Science* 1999;284(5421):1835-7.
[185] Cella M, Jarrossay D, Facchetti F, Alebardi O, Nakajima H, Lanzavecchia A, et al. Plasmacytoid monocytes migrate to inflamed lymph nodes and produce large amounts of type I interferon [see comments]. *Nat. Med.* 1999;5(8):919-23.
[186] Chung E, Amrute SB, Abel K, Gupta G, Wang Y, Miller CJ, et al. Characterization of virus-responsive plasmacytoid dendritic cells in the rhesus macaque. *Clin. Diagn. Lab. Immunol.* 2005;12(3):426-35.
[187] Coates PT, Barratt-Boyes SM, Zhang L, Donnenberg VS, O'Connell PJ, Logar AJ, et al. Dendritic cell subsets in blood and lymphoid tissue of rhesus monkeys and their mobilization with Flt3 ligand. *Blood* 2003;102(7):2513-21.
[188] Teleshova N, Jones J, Kenney J, Purcell J, Bohm R, Gettie A, et al. Short-term Flt3L treatment effectively mobilizes functional macaque dendritic cells. *J. Leukoc. Biol.* 2004;75(6):1102-10.
[189] Teleshova N, Kenney J, Jones J, Marshall J, Van Nest G, Dufour J, et al. CpG-C immunostimulatory oligodeoxyribonucleotide activation of plasmacytoid dendritic cells in rhesus macaques to augment the activation of IFN-gamma-secreting simian immunodeficiency virus-specific T cells. *J. Immunol.* 2004;173(3):1647-57.
[190] Ketloy C, Engering A, Srichairatanakul U, Limsalakpetch A, Yongvanitchit K, Pichyangkul S, et al. Expression and function of Toll-like receptors on dendritic cells and other antigen presenting cells from non-human primates. *Vet. Immunol. Immunopathol.* 2008;125(1-2):18-30.
[191] Kwissa M, Amara RR, Robinson HL, Moss B, Alkan S, Jabbar A, et al. Adjuvanting a DNA vaccine with a TLR9 ligand plus Flt3 ligand results in enhanced cellular immunity against the simian immunodeficiency virus. *J. Exp. Med.* 2007, 2007;204(11):2733-46.
[192] Medzhitov R. Recognition of microorganisms and activation of the immune response. *Nature* 2007;449(7164):819-26.
[193] Teleshova N, Kenney J, Williams V, Van Nest G, Marshall J, Lifson JD, et al. CpG-C ISS-ODN activation of blood-derived B cells from healthy and chronic immunodeficiency virus-infected macaques. *J. Leukoc. Biol.* 2006, 2006;79(2):257-67.
[194] Weiskopf D, Weinberger B, Grubeck-Loebenstein B. The aging of the immune system. *Transpl. Int.,* 2009;22(11):1041-50.
[195] Coe CL, Ershler WB. Intrinsic and environmental influences on immune senescence in the aged monkey. *Physiol. Behav.* 2001;73(3):379-84.
[196] Tarlinton D. B-cell memory: are subsets necessary? *Nat. Rev. Immunol.* 2006;6 (10):785-90.
[197] Bar-Or A, Oliveira EM, Anderson DE, Krieger JI, Duddy M, O'Connor KC, et al. Immunological memory: contribution of memory B cells expressing costimulatory molecules in the resting state. *J. Immunol.* 2001;167(10):5669-77.

[198] Weller S, Braun MC, Tan BK, Rosenwald A, Cordier C, Conley ME, et al. Human blood IgM "memory" B cells are circulating splenic marginal zone B cells harboring a prediversified immunoglobulin repertoire. *Blood* 2004;104(12):3647-54.

[199] Ermert K, Mitlohner H, Schempp W, Zachau HG. The immunoglobulin kappa locus of primates. *Genomics* 1995;25(3):623-9.

[200] Link JM, Hellinger MA, Schroeder HW, Jr. The rhesus monkey immunoglobulin IGHD and IGHJ germline repertoire. *Immunogenetics* 2002;54(4):240-50.

[201] Bible JM, Howard W, Robbins H, Dunn-Walters DK. IGHV1, IGHV5 and IGHV7 subgroup genes in the rhesus macaque. *Immunogenetics* 2003;54(12):867-73.

[202] Margolin DH, Saunders EH, Bronfin B, de Rosa N, Axthelm MK, Goloubeva OG, et al. Germinal center function in the spleen during simian HIV infection in rhesus monkeys. *J. Immunol.* 2006;177(2):1108-19.

[203] Cancro MP, Hao Y, Scholz JL, Riley RL, Frasca D, Dunn-Walters DK, et al. B cells and aging: molecules and mechanisms. *Trends Immunol.* 2009;30(7):313-8.

[204] Ademokun A, Wu YC, Dunn-Walters D. The ageing B cell population: composition and function. *Biogerontology* 2010;11(2):125-37.

[205] Frasca D, Landin AM, Riley RL, Blomberg BB. Mechanisms for decreased function of B cells in aged mice and humans. *J. Immunol.* 2008 1;180(5):2741-6.

[206] McKenna RW, Washington LT, Aquino DB, Picker LJ, Kroft SH. Immunophenotypic analysis of hematogones (B-lymphocyte precursors) in 662 consecutive bone marrow specimens by 4-color flow cytometry. *Blood* 2001;98(8):2498-507.

[207] Colonna-Romano G, Aquino A, Bulati M, Di Lorenzo G, Listi F, Vitello S, et al. Memory B cell subpopulations in the aged. *Rejuvenation Res.* 2006 Spring;9(1):149-52.

[208] Colonna-Romano G, Bulati M, Aquino A, Vitello S, Lio D, Candore G, et al. B cell immunosenescence in the elderly and in centenarians. *Rejuvenation Res.* 2008;11(2):433-9.

[209] Gibson KL, Wu YC, Barnett Y, Duggan O, Vaughan R, Kondeatis E, et al. B-cell diversity decreases in old age and is correlated with poor health status. *Aging Cell,* 2009;8(1):18-25.

[210] Chen ZW, Kou ZC, Shen L, Reimann KA, Letvin NL. Conserved T-cell receptor repertoire in simian immunodeficiency virus-infected rhesus monkeys. *J. Immunol.* 1993;151(4):2177-87.

[211] Jaeger EE, Bontrop RE, Lanchbury JS. Structure, diversity, and evolution of the T-cell receptor VB gene repertoire in primates. *Immunogenetics* 1994;40(3):184-91.

[212] Cheynier R, Henrichwark S, Wain-Hobson S. Sequence of the rhesus monkey T-cell receptor beta chain diversity and joining loci. *Immunogenetics* 1996;43(1-2):83-7.

[213] Levinson G, Hughes AL, Letvin NL. Sequence and diversity of rhesus monkey T-cell receptor beta chain genes. *Immunogenetics* 1992;35(2):75-88.

[214] Thiel C, Bontrop RE, Lanchbury JS. Structure and diversity of the T-cell receptor alpha chain in rhesus macaque and chimpanzee. *Hum. Immunol.* 1995;43(2):85-94.

[215] Miller JD, Masopust D, Wherry EJ, Kaech S, Silvestri G, Ahmed R. Differentiation of CD8 T cells in response to acute and chronic viral infections: implications for HIV vaccine development. *Curr. Drug Targets Infect. Disord.* 2005;5(2):121-9.

[216] Kannanganat S, Ibegbu C, Chennareddi L, Robinson HL, Amara RR. Multiple-cytokine-producing antiviral CD4 T cells are functionally superior to single-cytokine-producing cells. *J. Virol.* 2007;81(16):8468-76.

[217] Walker JM, Maecker HT, Maino VC, Picker LJ. Multicolor flow cytometric analysis in SIV-infected rhesus macaque. *Methods in cell biology* 2004;75:535-57.
[218] Pitcher CJ, Hagen SI, Walker JM, Lum R, Mitchell BL, Maino VC, et al. Development and homeostasis of T cell memory in rhesus macaque. *J. Immunol.* 2002;168(1):29-43.
[219] Hsu HC, Scott DK, Zhang P, Zhou J, Yang P, Wu Q, et al. CD8 T-cell immune phenotype of successful aging. *Mech. Ageing Develop.* 2006;127(3):231-9.
[220] Picker LJ, Reed-Inderbitzin EF, Hagen SI, Edgar JB, Hansen SG, Legasse A, et al. IL-15 induces CD4 effector memory T cell production and tissue emigration in nonhuman primates. *J. Clin. Invest.* 2006;116(6):1514-24.
[221] Surh CD, Sprent J. Homeostasis of Naive and Memory T Cells. *Immunity* 2008;29(6):848-62.
[222] Boyman O, Purton JF, Surh CD, Sprent J. Cytokines and T-cell homeostasis. *Curr. Opin. Immunol.* 2007;19(3):320-6.
[223] Purton JF, Tan JT, Rubinstein MP, Kim DM, Sprent J, Surh CD. Antiviral CD4+ memory T cells are IL-15 dependent. J. Exp. Med. 2007 April 16, 2007;204(4):951-61.
[224] Ansari AA, Mayne AE, Onlamoon N, Pattanapanyasat K, Mori K, Villinger F. Use of recombinant cytokines for optimized induction of antiviral immunity against SIV in the nonhuman primate model of human AIDS. *Immunol. Res.* 2004;29(1-3):1-18.
[225] Villinger F, Miller R, Mori K, Mayne AE, Bostik P, Sundstrom JB, et al. IL-15 is superior to IL-2 in the generation of long-lived antigen specific memory CD4 and CD8 T cells in rhesus macaques. *Vaccine* 2004;22(25-26):3510-21.
[226] Fagnoni FF, Vescovini R, Passeri G, Bologna G, Pedrazzoni M, Lavagetto G, et al. Shortage of circulating naive CD8(+) T cells provides new insights on immunodeficiency in aging. *Blood* 2000;95(9):2860-8.
[227] Naylor K, Li G, Vallejo AN, Lee WW, Koetz K, Bryl E, et al. The Influence of Age on T Cell Generation and TCR Diversity. *J. Immunol.* 2005;174(11):7446-52.
[228] Czesnikiewicz-Guzik M, Lee WW, Cui D, Hiruma Y, Lamar DL, Yang ZZ, et al. T cell subset-specific susceptibility to aging. *Clin. Immunol.* 2008;127(1):107-18
[229] Vallejo AN. CD28 extinction in human T cells: altered functions and the program of T-cell senescence. *Immunol. Rev.* 2005;205:158-69.
[230] Sansoni P, Vescovini R, Fagnoni F, Biasini C, Zanni F, Zanlari L, et al. The immune system in extreme longevity. *Exp. Gerontol.* 2008;43(2):61-5.
[231] Koch S, Larbi A, Ozcelik D, Solana R, Gouttefangeas C, Attig S, et al. Cytomegalovirus infection: a driving force in human T cell immunosenescence. *Ann. N Y Acad. Sci.* 2007;1114:23-35.
[232] Derhovanessian E, Larbi A, Pawelec G. Biomarkers of human immunosenescence: impact of Cytomegalovirus infection. *Curr. Opin. Immunol.* 2009;21(4):440-5.
[233] Trzonkowski P, Mysliwska J, Szmit E, Wieckiewicz J, Lukaszuk K, Brydak LB, et al. Association between cytomegalovirus infection, enhanced proinflammatory response and low level of anti-hemagglutinins during the anti-influenza vaccination--an impact of immunosenescence. *Vaccine* 2003;21(25-26):3826-36.
[234] Saurwein-Teissl M, Lung TL, Marx F, Gschosser C, Asch E, Blasko I, et al. Lack of antibody production following immunization in old age: association with CD8(+)CD28(-) T cell clonal expansions and an imbalance in the production of Th1 and Th2 cytokines. *J. Immunol.* 2002;168(11):5893-9.

[235] Vallejo AN, Weyand CM, Goronzy JJ. T-cell senescence: a culprit of immune abnormalities in chronic inflammation and persistent infection. *Trends Mol. Med.* 2004;10(3):119-24.
[236] Posnett DN, Sinha R, Kabak S, Russo C. Clonal populations of T cells in normal elderly humans: the T cell equivalent to "benign monoclonal gammapathy". *J. Exp. Med.* 1994;179(2):609-18.
[237] Schwab R, Szabo P, Manavalan JS, Weksler ME, Posnett DN, Pannetier C, et al. Expanded CD4+ and CD8+ T cell clones in elderly humans. *J. Immunol.* 1997;158(9):4493-9.
[238] Chen J. Senescence and functional failure in hematopoietic stem cells. Exp. Hematol. 2004;32(11):1025-32.
[239] Nikolich-Zugich J. Ageing and life-long maintenance of T-cell subsets in the face of latent persistent infections. *Nat. Rev. Immunol.* 2008;8(7):512-22.
[240] Jankovic V, Messaoudi I, Nikolich-Zugich J. Phenotypic and functional T-cell aging in rhesus macaques (Macaca mulatta): differential behavior of CD4 and CD8 subsets. *Blood* 2003;102(9):3244-51.
[241] Messaoudi I, Warner J, Nikolich-Zugich D, Fischer M, Nikolich-Zugich J. Molecular, cellular, and antigen requirements for development of age-associated T cell clonal expansions in vivo. *J. Immunol.* 2006;176(1):301-8.
[242] Messaoudi I, Warner J, Fischer M, Park B, Hill B, Mattison J, et al. Delay of T cell senescence by caloric restriction in aged long-lived nonhuman primates. *Proc. Natl. Acad. Sci. U S A,* 2006;103(51):19448-53.
[243] Cicin-Sain L, Messaoudi I, Park B, Currier N, Planer S, Fischer M, et al. Dramatic increase in naive T cell turnover is linked to loss of naive T cells from old primates. *Proc. Natl. Acad. Sci. U S A,* 2007;104(50):19960-5.
[244] Pawelec G, Larbi A, Derhovanessian E. Senescence of the human immune system. *J. Comp. Pathol.*;142 Suppl 1:S39-44.
[245] Wikby A, Ferguson F, Forsey R, Thompson J, Strindhall J, Lofgren S, et al. An immune risk phenotype, cognitive impairment, and survival in very late life: impact of allostatic load in Swedish octogenarian and nonagenarian humans. *J. Gerontol. A. Biol. Sci. Med. Sci.* 2005;60(5):556-65.
[246] Dykhuizen M, Ceman J, Mitchen J, Zayas M, MacDougall A, Helgeland J, et al. Importance of the CD3 marker for evaluating changes in rhesus macaque CD4/CD8 T-cell ratios. *Cytometry* 2000;40(1):69-75.
[247] Swack NS, Hsiung GD. Natural and experimental simian cytomegalovirus infections at a primate center. *J. Med. Primatol.* 1982;11(3):169-77.
[248] Vogel P, Weigler BJ, Kerr H, Hendrickx AG, Barry PA. Seroepidemiologic studies of cytomegalovirus infection in a breeding population of rhesus macaques. *Lab. Anim. Sci.* 1994;44(1):25-30.
[249] Bitmansour AD, Waldrop SL, Pitcher CJ, Khatamzas E, Kern F, Maino VC, et al. Clonotypic structure of the human CD4+ memory T cell response to cytomegalovirus. *J. Immunol.* 2001;167(3):1151-63.
[250] Price DA, Bitmansour AD, Edgar JB, Walker JM, Axthelm MK, Douek DC, et al. Induction and evolution of cytomegalovirus-specific CD4+ T cell clonotypes in rhesus macaques. *J. Immunol.* 2008;180(1):269-80.

[251] Hobbs MV, Ernst DN. T cell differentiation and cytokine expression in late life. *Developmental and comparative immunology* 1997;21(6):461-70.
[252] Gardner EM, Murasko DM. Age-related changes in Type 1 and Type 2 cytokine production in humans. *Biogerontology* 2002;3(5):271-90.
[253] O'Mahony L, Holland J, Jackson J, Feighery C, Hennessy TP, Mealy K. Quantitative intracellular cytokine measurement: age-related changes in proinflammatory cytokine production. *Clin. Exp. Immunol.* 1998;113(2):213-9.
[254] McNerlan SE, Rea IM, Alexander HD. A whole blood method for measurement of intracellular TNF-alpha, IFN-gamma and IL-2 expression in stimulated CD3+ lymphocytes: differences between young and elderly subjects. *Exp. Gerontol.* 2002;37(2-3):227-34.
[255] Sandmand M, Bruunsgaard H, Kemp K, Andersen-Ranberg K, Schroll M, Jeune B. High circulating levels of tumor necrosis factor-alpha in centenarians are not associated with increased production in T lymphocytes. *Gerontology* 2003;49(3):155-60.
[256] Zanni F, Vescovini R, Biasini C, Fagnoni F, Zanlari L, Telera A, et al. Marked increase with age of type 1 cytokines within memory and effector/cytotoxic CD8+ T cells in humans: a contribution to understand the relationship between inflammation and immunosenescence. *Exp. Gerontol.* 2003;38(9):981-7.
[257] Alberti S, Cevenini E, Ostan R, Capri M, Salvioli S, Bucci L, et al. Age-dependent modifications of Type 1 and Type 2 cytokines within virgin and memory CD4+ T cells in humans. *Mech. Ageing Develop.* 2006;127(6):560-6.
[258] Almanzar G, Schwaiger S, Jenewein B, Keller M, Herndler-Brandstetter D, Wurzner R, et al. Long-term cytomegalovirus infection leads to significant changes in the composition of the CD8+ T-cell repertoire, which may be the basis for an imbalance in the cytokine production profile in elderly persons. *J. Virol.* 2005;79(6):3675-83.
[259] Mascarucci P, Taub D, Saccani S, Paloma MA, Dawson H, Roth GS, et al. Age-related changes in cytokine production by leukocytes in rhesus monkeys. *Aging* (Milano) 2001;13(2):85-94.
[260] Mascarucci P, Taub D, Saccani S, Paloma MA, Dawson H, Roth GS, et al. Cytokine Responses in Young and Old Rhesus Monkeys: Effect of Caloric Restriction. *J. Interferon Cytokine* Res. 2002;22(5):565-71.
[261] Miller RA. Effect of aging on T lymphocyte activation. *Vaccine* 2000;18(16):1654-60.
[262] Nociari MM, Telford W, Russo C. Postthymic development of CD28-CD8+ T cell subset: age-associated expansion and shift from memory to naive phenotype. *J. Immunol.* 1999;162(6):3327-35.
[263] Valenzuela HF, Effros RB. Divergent telomerase and CD28 expression patterns in human CD4 and CD8 T cells following repeated encounters with the same antigenic stimulus. *Clin. Immunol.* 2002;105(2):117-25.
[264] Effros RB, Dagarag M, Spaulding C, Man J. The role of CD8+ T-cell replicative senescence in human aging. *Immunol. Rev.* 2005;205:147-57.
[265] Ershler WB, Coe CL, Laughlin N, Klopp RG, Gravenstein S, Roecker EB, et al. Aging and immunity in non-human primates. II. Lymphocyte response in thymosin treated middle-aged monkeys. *J. Gerontol.* 1988;43(5):B142-6.
[266] Taylor LD, Daniels CK, Schmucker DL. Ageing compromises gastrointestinal mucosal immune response in the rhesus monkey. *Immunology* 1992;75(4):614-8.

[267] Cicin-Sain L, Smyk-Pearson S, Currier N, Byrd L, Koudelka C, Robinson T, et al. Loss of naive T cells and repertoire constriction predict poor response to vaccination in old primates. *J. Immunol.*2010;184(12):6739-45.

[268] Stacy S, Pasquali A, Sexton VL, Cantwell AM, Kraig E, Dube PH. An age-old paradigm challenged: old baboons generate vigorous humoral immune responses to LcrV, a plague antigen. *J. Immunol.* 2008;181(1):109-15.

[269] Masoro EJ. Overview of caloric restriction and ageing. *Mech. Ageing Dev.* 2005;126(9):913-22.

[270] Masoro EJ. Subfield history: caloric restriction, slowing aging, and extending life. *Sci. Aging Knowledge Environ.* 2003;(8):RE2.

[271] Heilbronn LK, Ravussin E. Calorie restriction and aging: review of the literature and implications for studies in humans. *Am. J. Clin. Nutr.* 2003, 2003;78(3):361-9.

[272] Chen J, Astle CM, Harrison DE. Delayed immune aging in diet-restricted B6CBAT6 F1 mice is associated with preservation of naive T cells. *J. Gerontol. A. Biol. Sci. Med. Sci.* 1998;53(5):B330-7; discussion B8-9.

[273] Grossmann A, Maggio-Price L, Jinneman JC, Wolf NS, Rabinovitch PS. The effect of long-term caloric restriction on function of T-cell subsets in old mice. *Cell Immunol.* 1990;131(1):191-204.

[274] Spaulding CC, Walford RL, Effros RB. The accumulation of non-replicative, non-functional, senescent T cells with age is avoided in calorically restricted mice by an enhancement of T cell apoptosis. *Mech. Ageing Dev.* 1997;93(1-3):25-33.

[275] Effros RB, Walford RL, Weindruch R, Mitcheltree C. Influences of dietary restriction on immunity to influenza in aged mice. *J. Gerontol.* 1991;46(4):B142-7.

[276] Effros RB, Svoboda K, Walford RL. Influence of age and caloric restriction on macrophage IL-6 and TNF production. *Lymphokine Cytokine Res.* 1991;10(5):347-51.

[277] Spaulding CC, Walford RL, Effros RB. Calorie restriction inhibits the age-related dysregulation of the cytokines TNF-alpha and IL-6 in C3B10RF1 mice. *Mech. Ageing Dev.* 1997;93(1-3):87-94.

[278] Jolly CA. Dietary Restriction and Immune Function. *J. Nutr.* 2004 1;134(8):1853-6.

[279] Jolly CA, Fernandes G. Diet modulates Th-1 and Th-2 cytokine production in the peripheral blood of lupus-prone mice. *J. Clin. Immunol.* 1999;19(3):172-8.

[280] Weindruch R, Walford RL. Dietary restriction in mice beginning at 1 year of age: effect on life-span and spontaneous cancer incidence. *Science*;215(4538):1415-8.

[281] Volk MJ, Pugh TD, Kim M, Frith CH, Daynes RA, Ershler WB, et al. Dietary restriction from middle age attenuates age-associated lymphoma development and interleukin 6 dysregulation in C57BL/6 mice. *Cancer Res.* 1994;54(11):3054-61.

[282] Roberts SB, Schoeller DA. Human Caloric Restriction for Retardation of Aging: Current Approaches and Preliminary Data. *J. Nutr.* 2007;137(4):1076-7.

[283] Willcox BJ, Willcox DC, Todoriki H, Fujiyoshi A, Yano K, He Q, et al. Caloric restriction, the traditional okinawan diet, and healthy aging: the diet of the world's longest-lived people and its potential impact on morbidity and life span. *Ann. NY Acad. Sci.* 2007;1114(1):434-55.

[284] Lane MA, Ingram DK, Cutler RG, Knapka JJ, Barnard DE, Roth GS. Dietary restriction in nonhuman primates: progress report on the NIA study. *Ann. N Y Acad. Sci.* 1992;673:36-45.

[285] Ramsey JJ, Colman RJ, Binkley NC, Christensen JD, Gresl TA, Kemnitz JW, et al. Dietary restriction and aging in rhesus monkeys: the University of Wisconsin study. *Exp. Gerontol.* 2000;35(9-10):1131-49.
[286] Cefalu WT, Wagner JD, Bell-Farrow AD, Edwards IJ, Terry JG, Weindruch R, et al. Influence of caloric restriction on the development of atherosclerosis in nonhuman primates: progress to date. *Toxicol. Sci.* 1999;52(2 Suppl):49-55.
[287] Lane MA, Ingram DK, Roth GS. Calorie restriction in nonhuman primates: effects on diabetes and cardiovascular disease risk. *Toxicol. Sci.* 1999;52(2 Suppl):41-8.
[288] Messaoudi I, Fischer M, Warner J, Park B, Mattison J, Ingram DK, et al. Optimal window of caloric restriction onset limits its beneficial impact on T-cell senescence in primates. *Aging Cell,* 2008;7(6):908-19.
[289] Grossmann A, Rabinovitch PS, Lane MA, Jinneman JC, Ingram DK, Wolf NS, et al. Influence of age, sex, and dietary restriction on intracellular free calcium responses of CD4+ lymphocytes in rhesus monkeys (Macaca mulatta). *J. Cell Physiol.* 1995;162(2):298-303.
[290] Roecker EB, Kemnitz JW, Ershler WB, Weindruch R. Reduced immune responses in rhesus monkeys subjected to dietary restriction. *J. Gerontol. A. Biol. Sci. Med. Sci.* 1996;51(4):B276-9.
[291] Weindruch R, Lane MA, Ingram DK, Ershler WB, Roth GS. Dietary restriction in rhesus monkeys: lymphopenia and reduced mitogen-induced proliferation in peripheral blood mononuclear cells. *Aging* (Milano) 1997;9(4):304-8.
[292] Aspinall R, Mitchell W. Reversal of age-associated thymic atrophy: treatments, delivery, and side effects. *Exp. Gerontol.* 2008;43(7):700-5.
[293] Bradley LM, Haynes L, Swain SL. IL-7: maintaining T-cell memory and achieving homeostasis. *Trends Immunol.* 2005;26(3):172-6.
[294] Tan JT, Dudl E, LeRoy E, Murray R, Sprent J, Weinberg KI, et al. IL-7 is critical for homeostatic proliferation and survival of naive T cells. *Proc. Natl. Acad. Sci. U S A,* 2001;98(15):8732-7.
[295] Andrew D, Aspinall R. Il-7 and not stem cell factor reverses both the increase in apoptosis and the decline in thymopoiesis seen in aged mice. *J. Immunol.* 2001;166(3):1524-30.
[296] Beq S, Nugeyre MT, Ho Tsong Fang R, Gautier D, Legrand R, Schmitt N, et al. IL-7 induces immunological improvement in SIV-infected rhesus macaques under antiviral therapy. *J. Immunol.* 2006;176(2):914-22.
[297] Aspinall R, Pido-Lopez J, Imami N, Henson SM, Ngom PT, Morre M, et al. Old rhesus macaques treated with interleukin-7 show increased TREC levels and respond well to influenza vaccination. *Rejuvenation Res.* 2007;10(1):5-17.
[298] Douek DC, Koup RA. Evidence for thymic function in the elderly. *Vaccine* 2000;18(16):1638-41.
[299] Moniuszko M, Fry T, Tsai W-P, Morre M, Assouline B, Cortez P, et al. Recombinant Interleukin-7 Induces Proliferation of Naive Macaque CD4+ and CD8+ T Cells In Vivo. *J. Virol.* 2004 September 15, 2004;78(18):9740-9.
[300] Sportes C, Hakim FT, Memon SA, Zhang H, Chua KS, Brown MR, et al. Administration of rhIL-7 in humans increases in vivo TCR repertoire diversity by preferential expansion of naive T cell subsets. *J. Exp. Med.* 2008;205(7):1701-14.

[301] Min D, Panoskaltsis-Mortari A, Kuro OM, Hollander GA, Blazar BR, Weinberg KI. Sustained thymopoiesis and improvement in functional immunity induced by exogenous KGF administration in murine models of aging. *Blood* 2007;109(6):2529-37.
[302] Seggewiss R, Einsele H. Hematopoietic growth factors including keratinocyte growth factor in allogeneic and autologous stem cell transplantation. *Semin. Hematol.* 2007;44(3):203-11.
[303] Seggewiss R, Lore K, Guenaga FJ, Pittaluga S, Mattapallil J, Chow CK, et al. Keratinocyte growth factor augments immune reconstitution after autologous hematopoietic progenitor cell transplantation in rhesus macaques. *Blood* 2007 1;110(1):441-9.
[304] Ershler WB, Gravenstein S, Geloo ZS. Thymosin alpha 1 as an adjunct to influenza vaccination in the elderly: rationale and trial summaries. *Ann. N Y Acad. Sci.* 2007;1112:375-84.
[305] Urbanski HF, Noriega NC, Lemos DR, Kohama SG. Gene expression profiling in the rhesus macaque: experimental design considerations. *Methods* 2009;49(1):26-31.

In: Monkeys: Biology, Behavior and Disorders
Editor: Rachel M. Williams, pp 55-76

ISBN: 978-1-61209-911-8

Chapter 2

EFFECTS OF ADVERSE REARING EXPERIENCE ON ORGANIZATION OF BRAIN AND BEHAVIOR IN NONHUMAN PRIMATES

Bo Zhang and Eric E. Nelson
Section on Developmental Affective Neuroscience
National Institute of Mental Health
Bethesda MD 20892

Early adverse rearing experiments have been performed on nonhuman primates for over 60 years. These studies have generated important insights into the effects that the early social environment can exert on the developmental course of behavioral and physiological systems. Although a number of different models of adverse rearing have been used they all interrupt the interaction between infant and mother and result in abnormalities in similar domains. These include motor development, social behavior, emotional responsiveness, neuroendocrine dysregulation, alterations in central monoamne function, and abnormal development of some brain structures. We review these findings in this chapter.

Psychological development is a protracted process in humans, with brain circuits undergoing programmed maturational change well into the third decade of life (Huttenlocher and Dabholkar, 1997, Gogtay et al., 2008). However, the early infancy period appears to be a particularly formative time for the development of many social and emotional traits (Kreppner et al., 2007, Fox et al., 2010). One of the most important contributors to infant socio-emotional development across a number of mammalian species is interaction with the mother (Levine and Mody, 2003, Coplan et al., 2005, Suomi, 2005, Loman and Gunnar, 2010, Zhang and Meaney, 2010). Maternal behavior can usually be identified as a small set of behaviors such as nursing, carrying, protecting, retrieving, and grooming, directed by the mother toward the infant (Fleming et al., 1999). The manner in which these stimuli are administered to the infant can have long term consequences for both structural and functional maturation of brain circuits and for mental and physical health (Brown et al., 2009, Barrett and Fleming, 2010). Although there is a great deal of overlap in the types of nurturing behaviors directed toward the young across species, important differences exist as well, with more elaborate mother infant interactions apparent in highly social species such as many nonhuman primates, and more stereotyped patterns evident in animals with less elaborate social lives such as rodents. In addition one of the most important species differences in the effects of maternal care may not be in the type of maternal care administered but in the proportion of life in which maternal care is received. The period of dependency on the mother is much longer in most primates than most rodents and is even longer in humans than in other primates (Gibbons, 2008), thus the mother is likely to have a much greater impact on ultimate development of humans than any other species.

Although the types of maternal behavior do not vary much within species there is often a fair amount of inter-individual variability in the degree and the style with which these nurturant maternal behaviors are expressed. The effects of these variations have been the subject of many studies within the human developmental literature (Ainsworth, 1985, Field, 1996). Inter-individual variation in maternal behavior has also been the basis for elegant models of neurobiological development in rodents (Zhang and Meaney, 2010). Although there has been some attention in nonhuman primate models to naturally occurring variation in maternal behavior (Maestripieri et al., 2006), most of the studies in nonhuman primates have used more extreme and controlled manipulations of mother infant interaction imposed by the experimenter.

Existing studies on the effects of maternal behavior in humans, while important, are confounded by many factors beyond experimental control such as genetics, socio-economics, and indirect social experience among others. In contrast, while rodent models are elegant and well controlled, rodents have a markedly different neurobiology, social repertoire, and developmental trajectory than primates and so these models may be missing some important effects of mother infant interaction, particularly as they relate to neocortical development or the emergence of complex and flexible social behavior. Furthermore, there is over 90% overlap in the genome of human and nonhuman primates (Lovejoy, 1981), and many primates share highly similar neurobiology, social organization and patterns of development (Azmitia and Gannon, 1986, Bailey and Aunger, 1990, DeVore, 1990, Uylings and van Eden, 1990, Wright, 1990, Ebersberger et al., 2002, Fujiyama et al., 2002). Therefore the primate models of adverse rearing experience (ARE) offer a good intermediary between rodent and human experimental approaches, and particularly when combined with these other models, will likely provide important insights into basic principles of socio-emotional and neurobiological development that are affected by maternal care. Primate adverse rearing experiences (ARE) include isolate rearing, surrogate rearing, peer rearing, brief repeated separations and variable foraging demands. While these models can rightly be criticized for being overly austere and artificial, they also have many advantages. Subjects can easily be assigned randomly to different groups, developmental periods of manipulation can be tightly regulated and many other conditions (from social contact to room lighting) can be experimentally controlled. Moreover these manipulations often result in large group differences that are often needed in primate studies which are typically comprised of a very small number of subjects.

While there have been some recent studies on ARE effects in both new world and other old world primate genera (Levine and Mody, 2003, Marais et al., 2006, Law et al., 2009, Parker and Maestripieri, 2010), the overwhelming majority of primate ARE studies have been performed on macaques which is due in large part to the pioneering studies of Harry Harlow. Harlow and others characterized the social life of macaques as falling into three phases. The first phase which extends from birth until approximately 2 months of age is a time when infants spend virtually all of their time engaged in a ventral-ventral cling with their biological mothers. Harlow argued that during this time the infant develops specific and strong attachment bonds with the mother. The second phase which extends from approximately two to six months of age is a transitional time when infants begin to explore the physical and social environment in brief forays away from their mother, during this time infants spend increasing amounts of time participating social interactions especially playing with peers but maintain frequent contact with the mother. The third phase which extends from approximately 6 months of age until puberty (at 3-4 years of age), play with peers becomes the main social activity (Hinde and Spencer-Booth, 1967, Suomi, 1997, Suomi, 2005). In spite of the change in contact between infants and mothers, macaque juvenile monkeys always maintain a close social relationship with their mothers, and the mother continues to play the role of protector and a mentor to their offspring.

While behavioral observations indicate the importance of the mother for the daily organization of behavior for the macaque, the various ARE studies have demonstrated that disruption of mother-infant interaction produces many alterations in the biological and behavioral development of the organism. In this chapter we will begin by reviewing the various ARE methods that have been used to disrupt the normal mother-infant interaction in rhesus monkeys. We will then provide an overview of the types of effects these mother-infant disruptions have on behavioral and biological organization.

ARE Models

Three basic categories of rearing manipulation have been used: social isolation; peer rearing; and repeated maternal separations. In general all of these manipulations are initiated within days of birth and imposed for a period of several months after which monkeys are returned to more standard housing conditions so that the experimental period is restricted in time. This procedure allows investigators to focus in on specific sensitive periods of development.

Social Isolation

Social isolation was the first ARE approach adopted by Harlow in early 1960s. There are two kinds of social isolation: partial and total social isolation. In the total isolation condition the infant is removed from its mother immediately after birth and reared in a cage alone without any auditory, visual, olfactory or tactile contact from other monkeys including their mother (Harlow and Harlow, 1962a, Harlow et al., 1964, Harlow et al., 1965, Baysinger et al., 1972). In partial isolation, although infants are separately caged, there are other monkeys caged in the same room which allows infants to have visual, auditory and olfactory but not tactile contact with their conspecifics (Mason and Sponholz, 1963, Cross and Harlow, 1965, Suomi et al., 1971, Struble and Riesen, 1978). Although both total and partial isolation models produced robust differences in behavior, they have fallen out of favor because these manipulations produced profound cognitive and emotional deficits. More recent models have incorporated less profound manipulations of peer rearing and maternal separation and have generated less profoundly disturbed animals.

Surrogate /Peer Rearing

In surrogate mother rearing (SMR) infants are reared apart from other monkeys but are provided with an artificial "surrogate" mother (Harlow, 1958, Harlow and Zimmermann, 1959, Suomi, 1973, Kaplan, 1974, Eastman and Mason, 1975, Mason and Berkson, 1975, Roy et al., 1978, Hennessy and Kaplan, 1982, Schneider and Suomi, 1992, Capitanio and Mason, 2000, Dettmer et al., 2008). The surrogate mother typically consists of a soft, cloth covered cylinder with a spring so it is bouncy. The surrogate mother provides infants with both tactile and vestibular stimulation that resembles the mother. Infant monkeys spend a great deal of time in a ventral cling with their surrogates and use surrogates as a safe base from which to explore the environment just as they do with their mothers under natural conditions. SMR monkeys also form emotional attachments to their surrogates.

Peer-rearing (PR) (or nursery rearing, NR) is another widely used rearing condition during which the infant monkeys are reared together with a small group of peers of about the same age (Sackett, 1967, Chamove et al., 1973, Erwin et al., 1973, Worlein and Sackett, 1997). PR could be divided into continuous pair rearing, intermittent and rotational peer rearing. In continuous pair rearing condition infants are usually reared by pairs throughout development (Chamove et al., 1973, Novak and Sackett, 1997, Fekete et al., 2000, Hotchkiss and Paule, 2003). Intermittent peer rearing allows peer monkeys to contact with each other for a limited period of time and then infants are housed singly during the rest of the time (Rommeck et al., 2009). Within the rotational peer rearing condition, infants are continuously peer housed with different infant partners (Novak and Sackett, 1997, Rommeck et al., 2009). Thus the PR condition provides interaction with live and animate conspecifics but not the mother. Interestingly one characteristic of PR monkeys is that they tend to spend a great deal of time during the first year of life in mutual ventral clings, which is an uncommon occurrence between peers in mother reared monkeys of this age.

Surrogate-peer rearing (SPR) is a hybrid of SMR and PR conditions. For the first 3-6 months of life SPR infants are reared alone with an inanimate surrogate mother but allowed brief peer interactions for limited period of time during which the infants are put together just like the PR condition (Meyer et al., 1975, Bastian et al., 2003, Lutz et al., 2007). The SPR condition is the closer to natural rearing conditions than either PR or SMR because it provides peer experience (although this is limited) and provides opportunity for ventral cling and vestibular stimulation which is typically provided by the mother.

Maternal Separation

In contrast to the permanent removal of the mother in the methods described above, maternal separations allow the infant and mother to remain together during the entire period prior to weaning and receive natural maternal care. However forced separations between mother and infant are imposed. Although maternal separation is an emotionally aversive condition for infants, this procedure is not as encompassing as the permanent separation because infants continue to obtain maternal care throughout development, and generally produces less severe social and emotional disruptions, yet subtle developmental alterations remain. Furthermore maternal separation paradigms have more ecological validity for understanding human development as the separation conditions are more likely to resemble human experiences than are complete maternal separations which are rare, though not nonexistent in the human condition (Kreppner et al., 2007, Nelson et al., 2007, Pollak et al., 2010).

Early temporary separation studies involved a single removal of infants from their mothers although often for a prolonged period of days or weeks, followed by mother-infant reunion (Seay et al., 1962, Hinde et al., 1966, Kaufman and Rosenblum, 1967, Spencer-Booth and Hinde, 1971, Hinde and McGinnis, 1977). More recent studies have adopted brief repetitive mother-infant separations. In this model infants are removed from their natal group which includes mothers, and a small group of peers and other adults and placed in social isolation for relatively short periods of time (minutes or hours) and then returned to the group (Suomi et al., 1983, Clarke et al., 1998, Dettling et al., 2002, Levine and Mody, 2003, Sanchez et al., 2005). While these paradigms do not induce chronic deficits in maternal care, they do generate repeated experiences of distress, despair, and joy (after reunion). The impact of these procedures appears to be further intensified if the separations are unpredictable (Levine, 2000, Sanchez et al., 2005).

Another model which imposes brief maternal separations on the infant without actually removing the infant from its home environment is the variable foraging demand paradigm. In this paradigm, mothers are exposed to either a consistently low foraging demand (LFD) environment in which food is readily available and easily located; a high foraging demand (HFD) environment, in which mothers must consistently spend a great deal of time foraging for food; or a variable foraging demand (VFD) environment in which sometimes they are in a high demand and sometimes in a low demand environment (Rosenblum and Paully, 1984, Andrews and Rosenblum, 1991, Rosenblum and Andrews, 1994, Coplan et al., 1996). Importantly, this paradigm manipulates maternal behavior by changing the environment of the mother rather than the infant, and a key part of this paradigm is not only separation but also maternal stress and the predictability of the stress/separation exposure.

ARE Effects on Behavior

Locomotor Abnormalities

One of the most apparent effects of early total and partial isolate rearing studies was the abnormal pattern of motor behavior. Both partial and total isolated monkeys showed marked increases in odd motor behaviors including crouching, rocking, pacing, flipping, thumb-sucking and miscellaneous stereotypies (Harlow and Harlow, 1962a, Mason and Sponholz, 1963, Mitchell, 1968, Harlow and Suomi, 1971a, Suomi et al., 1971). They also engaged in more self-directed behavior such as self-clasping, self-manipulation, scratching, eye-pocking, self-grasping, self-rubbing, and autoeroticism (Baysinger et al., 1972). Often the self-directed behaviors become aggressive and include self-biting, hair pulling, and other self-injurious behaviors (SIB) with males showing a much higher level of vulnerability for SIB than females in adolescence and adulthood (Cross and Harlow, 1965, Suomi et al., 1971). Many of these same abnormal motor behaviors are also present in PR and SMR monkeys although they tend not to be as severe and there are some subtle differences among the various AREs. PR monkeys tend to show more ventral cling than SMR monkeys while SMR monkeys engage in more self-directed behaviors and SIBs (Suomi et al., 1971, Lutz et al., 2003). Abnormalities in motor behavior have not generally been reported following repeated separation or VFD rearing paradigms (Rosenblum et al., 2001).

Another motor abnormality that has been observed following isolate, peer, and surrogate rearing is difficulty with coordinating the motor patterns for sexual behavior. ARE reared males do not mount properly as they engage in varied but misplaced heterosexual efforts, while females do not maintain the sexual present (stands quadripedally with the perineal area directed towards the recipient), and when mounting attempts are made, the females often will fall down or turn their bodies (Wallen et al., 1981, Goldfoot et al., 1984). While PR and SMR monkeys generally can execute reproductive behavior, albeit in a clumsy fashion, appropriate sexual behaviors are virtually absent among isolate-reared monkeys when they reach puberty and adulthood (Harlow, 1962, Harlow et al., 1966).

Motivational Disturbances

All ARE paradigms (including VFD and repeated separations) have been shown to produce differences in primary emotional behavior that persists well beyond the period of manipulation. Indeed the persistent effects of ARE manipulations on emotional responding of mature individuals is one aspect of adverse rearing models that make them particularly attractive for studies that focus on sensitive periods of development. Adversely reared monkeys show increases in fear as expressed by enhanced expression of fear related behavior in the face of a threat, increased inhibitory responding in the face of mild threats, and elevations in fear potentiated startle, for example (Champoux, Metz et al. 1991; Suomi 1991; Rosenblum, Forger et al. 2001; Sanchez, Noble et al. 2005). Interestingly adversely reared monkeys also tend to show increases in reward related behavior as well. For example, PR monkeys consume more of a palatable solution in their home cage and VFD monkeys have been found to gain more weight and be more susceptible to insulin resistance in adolescence when given ad libitum access to food in their home environment (Mitchell, 1968, Chamove et al., 1973, Kaufman et al., 2007). Other studies have found that adult peer reared monkeys display both polyphagia and polydipsia in their home cage (Miller et al., 1969). A number of studies have also found that PR monkeys consume significantly more alcohol than MR monkeys when put in a "happy hour" type of environment (Higley, Hasert et al. 1991; Fahlke, Lorenz et al. 2000). On the other hand many studies have reported a blunting of emotional responding in ARE monkeys. For example adverse rearing has been found to reduce levels of environmental and social exploration (Mason and Sponholz, 1963, Griffin and Harlow, 1966, Mitchell, 1968, Rosenblum and Paully, 1984, Ruppenthal et al., 1991) and to reduce interest in maternal and sexual behaviors (Harlow et al., 1965, Suomi, 1978). The exact conditions under which some motivated behaviors are enhanced and others are blunted in ARE animals is not clear and is an important area for future research.

Social Deficiency

A number of alterations in social behavior have been consistently found in adversely reared monkeys. During infancy and the early juvenile phase ARE monkeys display a dramatic increase in ventral-ventral clinging behavior with peers (Chamove et al., 1973). Since one of the major behaviors that infant macaque monkeys engage in with the mother is ventral-ventral cling, this may be a compensatory response. Thus the ventral-ventral cling may serve an important regulatory function during infancy.

As ARE monkeys get older the excessive clinginess often turns into a social withdrawal, especially when encountering unfamiliar individuals. Juvenile ARE monkeys engage in less social play (Harlow et al., 1965). As adults ARE monkeys also engage in less affiliative behaviors such as grooming, side by side sitting and following than normally reared monkeys (Kraemer and McKinney, 1979). Adversely reared monkeys also have particular difficulty in adapting to new social groups or novel social environments, suggesting a general deficit in social flexibility (Harlow and Harlow, 1962b, Mason and Sponholz, 1963, Griffin and Harlow, 1966, Ruppenthal et al., 1991). A number of studies have also reported disruptions in maternal behavior in ARE females. Increases in neglect, abuse, and generally less competent maternal care have been observed (Seay, Alexander et al. 1964; Harlow and Suomi 1971; Suomi, Harlow et al. 1974; Suomi 1978; Suomi and Ripp 1983; Champoux, Byrne et al. 1992; Bridges, Slais et al. 2008). These findings underscore the transgenerational nature of parental behavior in a number of species and suggest that some aspects of maternal behavior may be learned during infancy (Fleming et al., 2002, Maestripieri, 2005).

Consistent differences have also been reported in aggressive behavior of ARE monkeys. Although aggressive behaviors are generally quite uncommon in infancy it seems to be especially rare in ARE monkeys (Harlow et al., 1965, Chamove et al., 1973). However, as ARE animals enter adolescence and adulthood marked increases in aggression occur. Increased aggressive behavior is directed at peers (Chamove, Rosenblum et al. 1973; Winslow, Noble et al. 2003), infants (Mitchell, 1968, Suomi et al., 1974) and adults (Chamove et al., 1973). Infant and adult directed aggression is quite uncommon in normally reared monkeys.

In spite of the fact that ARE generally increases aggression, a number of studies have found that ARE results in a reduction in dominance rank (Farrington and Loeber, 2000, Kessler, 2003). In macaques dominance is a complicated phenomenon that depends on a number of factors including maternal rank, age, sex, physical appearance, and affiliative behaviors in addition to appropriate aggressive behavior. Differences in early rearing experiences could result in lower dominance rank as a result of abnormal agonistic and affiliative behaviors among other factors including no clear association with a matriline.

Defects in Learning and Memory

Although not extensively investigated, some studies have found cognitive deficits in ARE adult monkeys. While adversely reared primates seem to be intact on simple discrimination and working memory tasks (Gluck et al., 1973), they show impairments in tasks that are more complex, such as those that require engaging working memory with dynamic rules or delays (Gluck, Harlow et al. 1973; Gluck and Sackett 1976; Beauchamp and Gluck 1988; Beauchamp, Gluck et al. 1991; Sanchez, Hearn et al. 1998). ARE monkeys also show impairments on tasks that require inhibition or response reversal (Gluck and Sackett, 1976, Sanchez et al., 1998). One study reported that even brief social isolation could impair performance in a multiple video-task assessment in adult rhesus monkeys (Washburn and

Rumbaugh, 1991). These results are consistent with several recent human studies that found early institutionalization or neglect impaired performance relative to age-matched control subjects on a number of cognitive and memory tasks (Nelson et al., 2007, Bauer et al., 2009, Majer et al., 2010). These studies have shown that although early life is a sensitive period for cognitive development, the period in which nurturance influences general cognitive development continues through at least the early juvenile period (Nelson et al., 2007).

Neurobiological Abnormalities

Dysregulation of the HPA Axis

One of the most consistent findings from ARE studies is a dysregulation of the hypothalamo-pitutiary-adrenal (HPA) axis, although like the motivational findings reported above the direction of the dysregulattion is not always consistent (Nelson and Winslow, 2009). A number of studies have found hyper-cortisolism under basal conditions in juvenile and adult ARE monkeys (Champoux et al., 1989, Suomi, 1991, Barrett et al., 2009). Increased levels of adrenocorticotrophin hormone (ACTH) and corticotropin releasing factor (CRF) have also been found in the plasma of ARE monkeys under basal conditions. Increased levels of CRF have also been found in cerebrospinal fluid in VFD monkeys (Coplan, Andrews et al. 1996), and increases the density of CRH binding sites have also been reported in many brain regions including prefrontal cortex, amygdala and hippocampus (Anisman et al., 1998) suggesting HPA changes may not be restricted to the periphery. On the other hand a number of studies have also reported hypo-cortisolism as a result of ARE. Lower basal levels of both ACTH and cortisol have been reported in a number of studies in ARE monkeys (Clarke et al., 1998, Shannon et al., 1998, Bartesaghi and Severi, 2004, Capitanio et al., 2005, Sanchez, 2006), while a few studies have even found no difference (Champoux et al., 1989, Clarke, 1993, Shansky and Morrison, 2009).

Similar results have also been found with the HPA response to stress. While many studies have found a potentiated HPA reaction and a prolonged elevation of cortisol following a stressful experience in ARE monkeys (Suomi, 1991, Higley et al., 1992, Dettling et al., 1998, Fahlke et al., 2000, Lyons et al., 2000, Sanchez et al., 2005), others have found a blunted response (Meyer et al., 1975, Clarke, 1993, Barr et al., 2004). In humans likewise, both hyper (Kaufman et al., 1997, Gunnar et al., 2001, Essex et al., 2002) and hypo-cortisolism (De Bellis et al., 1994, Hart J, 1995, Heim et al., 2000, Gunnar and Vazquez, 2001, Dozier et al., 2006, Elzinga et al., 2008, Bruce et al., 2009, Carpenter et al., 2009) have been reported following early life adversity.

A number of studies have also found disruptions in the circadian rhythm of cortisol secretion following ARE. Increased morning cortisol, diurnal flattening, and phase delays have all been reported (Boyce et al., 1995, Sanchez et al., 2005, Sanchez, 2006, Barrett et al., 2009, Nelson and Winslow, 2009). Similar circadian abnormalities have also been found in humans with the most consistent being an elevated morning cortisol response (Ruppenthal et al., 1991, Gonzalez et al., 2009, Gustafsson et al., 2010), although other abnormalities have also been reported (Carlson and Earls, 1997, Gunnar and Vazquez, 2001, Dozier et al., 2006, Cicchetti et al., 2010).

A few studies have also found changes in the sleep-wake patterns of ARE monkeys, further suggesting a fundamental disruption in circadian organization of behavior. Sleep changes include a decrease in total sleep time, reduced levels of rapid eye movement (REM) sleep (Reite et al., 1974, Reite and Short, 1977, Reite and Short, 1978), earlier morning waking and longer periods of sleep during the active period especially in the middle morning (Barrett et al., 2009) in ARE monkeys.

All of these studies suggest that the HPA axis is sensitive to early life experiences but the reason for different types of dysregulation is not entirely clear. However the HPA axis is clearly a dynamic system with many points of control. Cortisol secretion from the adrenal gland is induced by ACTH release from the pituitary which is controlled by CRF and AVP secretion in hypothalamus, which itself is under negative feedback control from both hypothalamic and hippocampal projections. Dysregulation in any one of these control points may lead to compensatory changes in others and hence unpredictable direction of change. The nature of the dysregualtion may be attributable to factors intrinsic to the individual or to aspects of the stressful experience that have yet to be discovered. A better understanding of this process awaits future research.

Abnormal Monoamine Function

Adverse rearing has been consistently associated with reduced functioning of brain serotonin. Both peer rearing and naturally occurring maternal rejection is associated with significantly reduced CSF levels of the serotonin metabolite 5-HIAA (Higley et al., 1996, Fahlke et al., 2000, Shannon et al., 2005, Maestripieri et al., 2006). VFD reared monkeys also appear to have alterations in serotonergic functions as they are hyporesponsive to the serotonergic probe, mCPP (Rosenblum et al., 1994). Interestingly a recent study showed that only receipt of aggression from the mother, but not from other social group members, is associated with lower post-stressor expression of the serotonin transporter (SERT), suggesting that there might be something specific about the mother-infant relationship that impacts serotonergic systems (Kinnally, Tarara et al. 2010). Furthermore PET studies in mature PR monkeys revealed reduced binding potential for SERT and for the 5HT1A receptor across a range of brain areas (Ichise, Vines et al. 2006; Spinelli, Chefer et al. 2010), suggesting the effects of ARE on serotonin function may be both chronic and widespread. Finally there is some indication that selective serotonin reuptake inhibitors may ameliorate some of the effects of ARE (Higley et al., 1998).

Consistent with serotonin playing a central role in adversity associated with ARE, a number of recent studies have found an interaction between polymorphisms of the 5-HTT gene or the monoamine oxidase gene (MAOA) and the severity of the effects induced by ARE. ARE and 5HTT polymorphisms interact in alcohol consumption (Barr, Newman et al. 2004), HPA axis regulation (Barr, Newman et al. 2004), the intensity of the separation response (Spinelli, Schwandt et al. 2007) and the functioning of the serotonin system itself (Bennett, Lesch et al. 2002). While the MAOA gene polymorphisms have been found to interact with ARE in the expression of aggressive behavior (Barr, Newman et al. 2004; Newman, Syagailo et al. 2005). Importantly, the interaction between adversity and serotonin transporter polymorphisms has also been linked to psychiatric disorder in humans (Caspi, Sugden et al. 2003; Kumsta, Stevens et al. 2010; Nordquist and Oreland 2010; Spinelli,

Chefer et al. 2010). Interestingly in both the primate models and in human studies the adversity by serotonin interaction appears stronger in females than males (Barr, Newman et al. 2004; Wust, Kumsta et al. 2009; Hammen, Brennan et al. 2010). This sex specific relationship may provide insight into why females are twice as likely to develop mood and anxiety disorders as males.

In addition to serotonin, catecholamine function has also been found to be altered in ARE animals. PR reared infant monkeys were found to have significantly lower CSF concentrations of norepinephrine, and the catecholamine metabolite homovanillic acid (HVA) (Kraemer et al., 1989). While a similar study found elevated levels of norepinephrine in CSF and this was accompanied by lower levels of the metabolites HVA and DOPAC indicating lower turnover of catecholeamines in brain (Clarke, Hedeker et al. 1996; Clarke, Ebert et al. 1999). PR monkeys have also been found to have attenuated NE secretion during a social separation manipulation suggesting chronically compromised function (Clarke, Hedeker et al. 1996; Clarke, Ebert et al. 1999). Furthermore, VFD reared monkeys are hyper-responsive to the noradrenergic probe yohimbine (Rosenblum et al., 1994).

Other Neurochemical Abnormalities

Although they have not received as much attention in the literature a number of neuropeptides are likely to be affected by ARE in primates (Panksepp, Nelson et al. 1997; Nelson and Panksepp 1998). Oxytocin, vasopressin and endogenous opioids all have clear effects on affiliative behavior in a variety of mammalian species including both nonhuman primates and humans (Kalin, Shelton et al. 1988; Keverne, Martensz et al. 1989; Insel 2010; Smith, Agmo et al. 2010). One study has reported reduced levels of oxytocin in CSF of peer reared rhesus monkeys (Winslow et al., 2003), and these correlated with social deficits. However no study to our knowledge has assessed central opiod function in ARE animals nor has any study administered neuropeptide agents to ARE animals. Another neuropeptide which is likely affected by adverse rearing is CRF. CRF plays a role both as a releasing factor for ACTH as part of the HPA axis but also serves as a neurontransmiter in several regions thought to be involved in the affective response to fear and stress (Mathew, Price et al. 2008; Ohmura and Yoshioka 2009). Since there are clear effects of ARE on HPA axis functioning it seems likely that central CRF function may be altered as well. Some studies have reported ARE effects on CRF levels in spinal fluid (Coplan et al., 2005) and alterations in CRF receptor binding has been found to be altered in several brain regions following ARE (Anisman et al., 1998, Higley et al., 1998). Finally several other endocrine abnormalities have been found in ARE monkeys that are not directly linked with emotional responding. For example abberent responses of both insulin and growth hormone have been found in monkeys following VFD rearing (Coplan, Smith et al. 2000; Kaufman, Banerji et al. 2007).

Epigenetic Mechanism of ARE Effects

Emerging evidence largely from rodent studies indicates that epigenetic modifications may serve as a critical mechanism through which experiences encountered during early life may have sustained effects on behavior. Epigenetics refers to the modification of gene expression caused by mechanisms other than changes in the underlying DNA sequence. Differential interaction with the mother can induce localized changes to the genome such as methylation and acetylation of targeted genes which can have a large impact on the degree to which the inherited genes are actually expressed and these changes can have long lasting effects on physiological processes like HPA regulation (Champagne and Curley 2009; Bagot and Meaney 2010; Zhang and Meaney 2010). Recent studies have also found evidence for epigenetic regulation of serotonin systems in rhesus monkeys (Kumar, Sachar et al. 1985; Bohn, O'Banion et al. 1994). Many other physiological systems are also likely to be affected by epigenetic changes induced by maternal interaction (Sabatini, Ebert et al. 2007; Law, Pei et al. 2009; Law, Pei et al. 2009).

ARE Effects on Brain Structure and Function

Human studies have found structural alterations in a number of brain regions involved in affective responding following experience of early adversity. These include increases in volume of amygdala (Mehta, Golembo et al. 2009; Tottenham and Sheridan 2009; Tottenham, Hare et al. 2010); reductions in hippocampal volume (Bremner, Randall et al. 1997; Stein, Koverola et al. 1997; Cohen, Grieve et al. 2006; Woon and Hedges 2008; Tottenham and Sheridan 2009); reduced gray matter in orbitofrontal cortex (Hanson, Chung et al. 2010; Thomaes, Dorrepaal et al. 2010); increased volume of ventral prefrontal cortex (Richert et al., 2006); volume reductions in anterior cingulate (Thomas, Drevets et al. 2001; Cohen, Grieve et al. 2006; Kitayama, Quinn et al. 2006), cerebellum and corpus collosum (De Bellis, Baum et al. 1999; Teicher, Dumont et al. 2004; Bauer, Hanson et al. 2009).

However, there are only a few reports of volumetric or structural differences in nonhuman primate ARE models. There are no reports of amygdala volume in ARE primates and there are three reports of no difference in hippocampal volume following either PR or repeated separation rearing (Sanchez et al., 1998, Law et al., 2009, Spinelli et al., 2009). There is one report of increases in ventral prefrontal cortex and anterior cingulate and reductions in cerebellum (Spinelli, Chefer et al. 2009). There is also a separate report of reduced corpus colosum volume (Sanchez et al., 1998), but this was not found by Spinelli et al. (Spinelli, Chefer et al. 2009).

On the other hand there have been several findings of more specific neurochemical alterations within some of the regions not demonstrating volumetric changes. Several years ago Martin et al. found marked reductions in markers for substance P, leu-enkephalin, and catecholamines within the striatum of adult monkeys who had undergone ARE (Martin et al., 1991), and Siegel et al found increased expression of non-phosphorylated filament protein within the hippocampus of juvenile ARE monkeys suggesting greater degeneration of hippocampal neurons (Siegel et al., 1993). More recently, Law and colleagues found reduced expression of GAP-43, a protein associated with nerve growh, within the hippocampus of adult marmosets following a repeated separation paradigm. Reduced hippocampal levels of mRNA for the serotonin receptor 5HT1A and reduced 5HT1A binding was also found in this study (Law, Pei et al. 2009). In a subsequent study Law also reported reduced binding of 5HT1A receptor in anterior cingulate cortex of adversely reared adolescent marmosets (Law, Pei et al. 2009). Although there have been surprisingly few reports of abnormalities in the amygdala of ARE monkeys, Sabatini et al recently reported several genes in the amygdala which showed a different pattern of expression in 3 month old infants that had undergone maternal separation at different points in early life. One gene in particular (GUCY1A3) was also associated with behavioral differences between the different manipulations (Sabatini et al., 2007). In addition some studies have reported increased startle responses in ARE monkeys (Parr, Winslow et al. 2002; Nelson, Herman et al. 2009), a finding which suggests amygdala abnormalities (Antoniadis et al., 2009). Finally, some studies have found that neurons within the cerebellum, hippocampus, primary motor, somatosensory, and prefrontoal cortices of ARE monkeys have significantly fewer dentritic branches (Struble and Riesen 1978; Floeter and Greenough 1979; Stell and Riesen 1987; Kozorovitskiy, Gross et al. 2005) suggesting that adverse rearing may have widespread influence on network complexity within the central nervous system.

Summary and Future Directions

ARE studies over the last 60 years have demonstrated quite convincingly the dramatic influence that social experiences during the first few months of life exert on the ultimate organization of primate brain and behavior throughout life. These studies have led to important societal shifts in how we view the role of nurturance and emotional experience during early development (Blum, 2002). It is now taken as a given that early social experiences with mother affect the behavior of the organism by sculpting brain and endocrine systems during organizational periods. However there is a great deal that remains unknown and questions that remain to be investigated over the next 60 years.

For example while the first 6 months of life appear to be an important organizing period for many of the effects discussed above, this is a rather long period in developmental terms. Some studies have suggested that within this 6 month period different behaviors are organized at different periods of time (Blum, 2002, Sabatini et al., 2007). On the other hand sensitive periods of brain and behavior organization are likely not limited to early life but extend at least until puberty (Nelson, Leibenluft et al. 2005; Andersen and Teicher 2008; Gogtay, Lu et al. 2008; Forbes and Dahl 2010). Furthermore there are likely to be many effects of ARE that are latent or quiescent for a number of years and only apparent in puberty or adulthood (Tottenham and Sheridan, 2009). Thus future ARE models of development would benefit from both a more micro and a more macro approach to development.

A second issue with most current ARE models is the severity of the manipulation. Rearing in complete absence of the mother is an extreme condition not typically encountered in nature by either humans or animals. Thus the extent to which development under these conditions informs development under less austere conditions is unclear. Although more recent paradigms such as repeated separations or variable foraging demand have found other ways of manipulating the mother-infant relationship without removing the mother entirely, there remains a lack of ecological validity. In order to truly relate to human condition it seems that more naturalistic paradigms should be developed. This has been done in some NHP studies (Maestripieri et al., 2006), although on a relatively limited basis. Rodent models which have identified extremes of normal behavior and used cross fostering methods have produced very compelling models of the effects of normally occurring stylistic variation in maternal behavior (Zhang and Meaney, 2010). Future ARE paradigms could benefit from such models.

Another issue that is beginning to emerge in some ARE studies is the effect of gender. Gender is an extremely important factor in most psychiatric conditions (Giedd and Rapoport, 2010) and understanding how developmental trajectories following adverse experiences might differ in males and females will likely be highly relevant issue for translational research. In general females demonstrate a preponderance of internalizing disorders like anxiety and depression, whereas males predominate in externalizing disorders such as those that involve aggression (Farrington and Loeber 2000; Kessler 2003). A similar trend has been reported in ARE models with males more prone to aggression and self-injurious behavior (Cross and Harlow 1965; Suomi, Harlow et al. 1971; Rommeck, Anderson et al. 2009), and reduced dominance rank (Mitchell, 1968). While female ARE monkeys are more likely to show alterations in HPA axis and excessive alcohol consumption (Barr, Newman et al. 2004; Sanchez, Noble et al. 2005). Uncovering the developmental or neurobiological reason for these gender differences would likely be of great interest to the psychiatric community.

Finally, an issue that was of interest many years ago but which seems to have received less attention recently is rehabilitation. Early conceptions of critical periods of organization have generally given way to the concept of sensitive periods of organization (Lewis and Maurer, 2005). While specific brain regions are undergoing programmed synapse organization and myelination they are particularly sensitive to environmental experience though plasticity in these regions continues to be possible throughout life. However there are very few connections which are absolute and most are malleable although it takes more effort and stronger input outside the sensitive period than inside. In other words you can teach an old dog a new trick it just might take a very long time. In the 1970s, Suomi conducted a series of studies demonstrating that ARE monkeys could undergo successful "therapy" by being exposed to the right social environment. After pairing with "therapist" monkeys, isolated infants show decline of self-clasping, huddling, rock, stereotypic behavior and increase of social contact and of social play (Harlow and Suomi, 1971b). Surrogates could partially rehabilitate the defects of isolate-reared monkeys as they exhibited significant recovery of basic social and nonsocial behavior in terms of increases in locomotive, exploratory, and contact-oriented social responses patterns and significant declines in self-directed disturbance activities (Suomi, 1973). More recently Higley reported that antidepressant medications could reduce the excessive alcohol intake often observed in PR monkeys (Higley et al., 1998). These kinds of studies provide invaluable information to the psychiatric and psychological community in terms of the potential of different therapeutic approaches for humans with mood or anxiety disorders who come from adverse rearing circumstances. Unfortunately only a few such studies have been published with this approach and it should be a greater area of focus in future studies.

In summary, nonhuman primate adverse rearing paradigms have been extensively used over the past half century and have contributed tremendously to our appreciation of the importance of the early rearing environment. Although there are a number of variations on ARE paradigms all have demonstrated that manipulation of the mother-infant environment has a major effect on ultimate behavioral, neural, and endocrine organization. Although much has been learned from studies to date, there is much more to understand about early environmental impact on developmental trajectory and future ARE studies will no doubt add further clarity.

REFERENCES

Ainsworth MD (Patterns of infant-mother attachments: antecedents and effects on development. *Bull. N.Y. Acad. Med.* 61:771-791.1985).

Andrews MW, Rosenblum LA (Attachment in monkey infants raised in variable- and low-demand environments. *Child Dev.* 62:686-693.1991).

Anisman H, Zaharia MD, Meaney MJ, Merali Z (Do early-life events permanently alter behavioral and hormonal responses to stressors? *Int. J. Dev. Neurosci.* 16:149-164.1998).

Antoniadis EA, Winslow JT, Davis M, Amaral DG (The nonhuman primate amygdala is necessary for the acquisition but not the retention of fear-potentiated startle. *Biol. Psychiatry,* 65:241-248.2009).

Azmitia EC, Gannon PJ (The primate serotonergic system: a review of human and animal studies and a report on Macaca fascicularis. *Adv. Neurol.* 43:407-468.1986).

Bailey R, Aunger R (Humans as primates: The social relationships of Efe pygmy men in comparative perspective. *International Journal of Primatology* 11:127-146.1990).

Barr CS, Newman TK, Shannon C, Parker C, Dvoskin RL, Becker ML, Schwandt M, Champoux M, Lesch KP, Goldman D, Suomi SJ, Higley JD (Rearing condition and rh5-HTTLPR interact to influence limbic-hypothalamic-pituitary-adrenal axis response to stress in infant macaques. *Biol. Psychiatry,* 55:733-738.2004).

Barrett CE, Noble P, Hanson E, Pine DS, Winslow JT, Nelson EE (Early adverse rearing experiences alter sleep-wake patterns and plasma cortisol levels in juvenile rhesus monkeys. *Psychoneuroendocrinology* 34:1029-1040.2009).

Barrett J, Fleming AS (Annual Research Review: All mothers are not created equal: neural and psychobiological perspectives on mothering and the importance of individual differences. *J. Child Psychol. Psychiatry*. 2010).

Bartesaghi R, Severi S (Effects of early environment on field CA2 pyramidal neurons in the guinea-pig. *Neuroscience* 123:703-714.2004).

Bastian ML, Sponberg AC, Suomi SJ, Higley JD (Long-term effects of infant rearing condition on the acquisition of dominance rank in juvenile and adult rhesus macaques (Macaca mulatta). *Dev. Psychobiol.* 42:44-51.2003).

Bauer PM, Hanson JL, Pierson RK, Davidson RJ, Pollak SD (Cerebellar volume and cognitive functioning in children who experienced early deprivation. *Biol. Psychiatry,* 66:1100-1106.2009).

Baysinger CM, Brandt EM, Mitchell G (Development of infant social isolate monkeys (Macaca mulatta) in their isolation environments. Primates 13:257-270.1972).

Blum D (2002) Love at Goon Park. New York: Berkley Publishing.

Boyce WT, Champoux M, Suomi SJ, Gunnar MR (Salivary cortisol in nursery-reared rhesus monkeys: reactivity to peer interactions and altered circadian activity. *Dev. Psychobiol.* 28:257-267.1995).

Brown DW, Anda RF, Tiemeier H, Felitti VJ, Edwards VJ, Croft JB, Giles WH (Adverse childhood experiences and the risk of premature mortality. *Am. J. Prev. Med.* 37:389-396.2009).

Bruce J, Fisher PA, Pears KC, Levine S (Morning cortisol levels in preschool-aged foster children: Differential effects of maltreatment type. *Developmental. Psychobiology,* 51:14-23.2009).

Capitanio JP, Mason WA (Cognitive style: problem solving by rhesus macaques (Macaca mulatta) reared with living or inanimate substitute mothers. *J. Comp. Psychol.* 114:115-125.2000).

Capitanio JP, Mendoza SP, Mason WA, Maninger N (Rearing environment and hypothalamic-pituitary-adrenal regulation in young rhesus monkeys (Macaca mulatta). *Dev. Psychobiol.* 46:318-330.2005).

Carlson M, Earls F (Psychological and neuroendocrinological sequelae of early social deprivation in institutionalized children in Romania. *Ann. N.Y. Acad. Sci.* 807:419-428.1997).

Carpenter LL, Tyrka AR, Ross NS, Khoury L, Anderson GM, Price LH (Effect of childhood emotional abuse and age on cortisol responsivity in adulthood. *Biol. Psychiatry,* 66:69-75.2009).

Chamove AS, Rosenblum LA, Harlow HF (Monkeys (Macaca mulatta) raised only with peers. A pilot study. *Anim. Behav.* 21:316-325.1973).

Champoux M, Coe CL, Schanberg SM, Kuhn CM, Suomi SJ (Hormonal effects of early rearing conditions in the infant rhesus monkey. *American Journal of Primatology,* 19:111-117.1989).

Cicchetti D, Rogosch FA, Gunnar MR, Toth SL (The differential impacts of early physical and sexual abuse and internalizing problems on daytime cortisol rhythm in school-aged children. *Child Development* 81:252-269.2010).

Clarke AS (Social rearing effects on HPA axis activity over early development and in response to stress in rhesus monkeys. *Dev. Psychobiol.* 26:433-446.1993).

Clarke AS, Kraemer GW, Kupfer DJ (Effects of rearing condition on HPA axis response to fluoxetine and desipramine treatment over repeated social separations in young rhesus monkeys. *Psychiatry Res.* 79:91-104.1998).

Coplan JD, Altemus M, Mathew SJ, Smith EL, Sharf B, Coplan PM, Kral JG, Gorman JM, Owens MJ, Nemeroff CB, Rosenblum LA (Synchronized maternal-infant elevations of primate CSF CRF concentrations in response to variable foraging demand. CNS Spectr 10:530-536.2005).

Coplan JD, Andrews MW, Rosenblum LA, Owens MJ, Friedman S, Gorman JM, Nemeroff CB (Persistent elevations of cerebrospinal fluid concentrations of corticotropin-releasing factor in adult nonhuman primates exposed to early-life stressors: implications for the pathophysiology of mood and anxiety disorders. *Proc. Natl. Acad. Sci. U S A,* 93:1619-1623.1996).

Cross HA, Harlow HF (Prolonged and progressive effects of partial isolation on the behavior of macaque monkeys. *J. exp. Res. Pers.* 1:39-49.1965).

De Bellis MD, Chrousos GP, Dorn LD, Burke L, Helmers K, Kling MA, Trickett PK, Putnam FW (Hypothalamic-pituitary-adrenal axis dysregulation in sexually abused girls. *J. Clin. Endocrinol. Metab.* 78:249-255.1994).

Dettling A, Pryce CR, Martin RD, Dobeli M (Physiological Responses to parental separation and a strange situation are related to parental care received in juvenile Goeldi's monkeys (Callimico goeldii). *Dev. Psychobiol.* 33:21-31.1998).

Dettling AC, Feldon J, Pryce CR (Repeated parental deprivation in the infant common marmoset (Callithrix jacchus, primates) and analysis of its effects on early development. *Biol. Psychiatry,* 52:1037-1046.2002).

Dettmer AM, Ruggiero AM, Novak MA, Meyer JS, Suomi SJ (Surrogate mobility and orientation affect the early neurobehavioral development of infant rhesus macaques (Macaca mulatta). *Dev. Psychobiol.* 50:418-422.2008).

DeVore I (Introduction: Current studies on primate socioecology and evolution. *International Journal of Primatology* 11:1-5.1990).

Dozier M, Manni M, Gordon MK, Peloso E, Gunnar MR, Stovall-McClough KC, Eldreth D, Levine S (Foster children's diurnal production of cortisol: an exploratory study. *Child Maltreat.* 11:189-197.2006).

Eastman RF, Mason WA (Looking behavior in monkeys raised with mobile and stationary artificial mothers. *Dev. Psychobiol.* 8:213-221.1975).

Ebersberger I, Metzler D, Schwarz C, Paabo S (Genomewide comparison of DNA sequences between humans and chimpanzees. *Am. J. Hum. Genet.* 70:1490-1497.2002).

Elzinga BM, Roelofs K, Tollenaar MS, Bakvis P, van Pelt J, Spinhoven P (Diminished cortisol responses to psychosocial stress associated with lifetime adverse events a study among healthy young subjects. *Psychoneuroendocrinology* 33:227-237.2008).

Erwin J, Mitchell G, Maple T (Abnormal behavior in non-isolate-reared rhesus monkeys. *Psychol. Rep.* 33:515-523.1973).

Essex MJ, Klein MH, Cho E, Kalin NH (Maternal stress beginning in infancy may sensitize children to later stress exposure: effects on cortisol and behavior. *Biol. Psychiatry,* 52:776-784.2002).

Fahlke C, Lorenz JG, Long J, Champoux M, Suomi SJ, Higley JD (Rearing experiences and stress-induced plasma cortisol as early risk factors for excessive alcohol consumption in nonhuman primates. *Alcohol. Clin. Exp. Res.* 24:644-650.2000).

Farrington DP, Loeber R (Epidemiology of juvenile violence. *Child Adolesc. Psychiatr. Clin. N. Am.* 9:733-748.2000).

Fekete JM, Norcross JL, Newman JD (Artificial turf foraging boards as environmental enrichment for pair-housed female squirrel monkeys. *Contemp. Top. Lab. Anim. Sci.* 39:22-26.2000).

Field T (Attachment and separation in young children. *Annu. Rev. Psychol.* 47:541-561.1996).

Fleming AS, Kraemer GW, Gonzalez A, Lovic V, Rees S, Melo A (Mothering begets mothering: the transmission of behavior and its neurobiology across generations. *Pharmacol. Biochem. Behav.* 73:61-75.2002).

Fleming AS, O'Day DH, Kraemer GW (Neurobiology of mother-infant interactions: experience and central nervous system plasticity across development and generations. *Neurosci. Biobehav. Rev.* 23:673-685.1999).

Fox SE, Levitt P, Nelson CA, 3rd (How the timing and quality of early experiences influence the development of brain architecture. *Child Dev.* 81:28-40.2010).

Fujiyama A, Watanabe H, Toyoda A, Taylor TD, Itoh T, Tsai SF, Park HS, Yaspo ML, Lehrach H, Chen Z, Fu G, Saitou N, Osoegawa K, de Jong PJ, Suto Y, Hattori M, Sakaki Y (Construction and analysis of a human-chimpanzee comparative clone map. *Science* 295:131-134.2002).

Gibbons A (Anthropology. The birth of childhood. *Science* 322:1040-1043.2008).

Giedd JN, Rapoport JL (Structural MRI of pediatric brain development: what have we learned and where are we going? *Neuron.* 67:728-734.2010).

Gluck JP, Harlow HF, Schiltz KA (Differential effect of early enrichment and deprivation on learning in the rhesus monkey (Macaca mulatta). *Journal of Comparative and Physiological Psychology*, 84:598-604.1973).

Gluck JP, Sackett GP (Extinction deficits in socially isolated rhesus monkeys (Macaca mulatta). *Developmental Psychology,* 12:173-174.1976).

Gogtay N, Lu A, Leow AD, Klunder AD, Lee AD, Chavez A, Greenstein D, Giedd JN, Toga AW, Rapoport JL, Thompson PM (Three-dimensional brain growth abnormalities in childhood-onset schizophrenia visualized by using tensor-based morphometry. *Proc. Natl. Acad. Sci. U S A,* 105:15979-15984.2008).

Goldfoot DA, Wallen K, Neff DA, McBrair MC, Goy RW (Social influence on the display of sexually dimorphic behavior in rhesus monkeys: isosexual rearing. *Arch. Sex Behav.* 13:395-412.1984).

Gonzalez A, Jenkins JM, Steiner M, Fleming AS (The relation between early life adversity, cortisol awakening response and diurnal salivary cortisol levels in postpartum women. *Psychoneuroendocrinology* 34:76-86.2009).

Griffin GA, Harlow HF (Effects of three months of total social deprivation on social adjustment and learning in the rhesus monkey. *Child Dev.* 37:533-547.1966).

Gunnar MR, Morison SJ, Chisholm K, Schuder M (Salivary cortisol levels in children adopted from romanian orphanages. *Dev. Psychopathol.* 13:611-628.2001).

Gunnar MR, Vazquez DM (Low cortisol and a flattening of expected daytime rhythm: potential indices of risk in human development. *Dev. Psychopathol.* 13:515-538.2001).

Gustafsson PE, Janlert U, Theorell T, Hammarström A (Life-course socioeconomic trajectories and diurnal cortisol regulation in adulthood. *Psychoneuroendocrinology* 35:613-623.2010).

Harlow H, Joslyn W, Senko M, Dopp A (Behavioral aspects of reproduction in primates. *Journal of animal science* 25:49.1966).

Harlow HF (The nature of love *American Psychologist,* 13:673-685.1958).

Harlow HF (The heterosexual affectional system in monkeys. *American Psychologist,* 17:1-9.1962).

Harlow HF, Dodsworth RO, Harlow MK (Total social isolation in monkeys. *Proc. Natl. Acad. Sci. U S A,* 54:90-97.1965).

Harlow HF, Harlow M (Social deprivation in monkeys. *Sci. Am.* 207:136-146.1962a).

Harlow HF, Harlow MK (The effect of rearing conditions on behavior. *Bull. Menninger Clin.* 26:213-224.1962b).

Harlow HF, Rowland GL, Griffin GA (The Effect of Total Social Deprivation on the Development of Monkey Behavior. *Psychiatr. Res. Rep. Am. Psychiatr. Assoc.* 19:116-135.1964).

Harlow HF, Suomi SJ (Production of depressive behaviors in young monkeys. *Journal of Autism and Developmental Disorders* 1:246-255.1971a).

Harlow HF, Suomi SJ (Social recovery by isolation-reared monkeys. *Proc. Natl. Acad. Sci. U S A,* 68:1534-1538.1971b).

Harlow HF, Zimmermann RR (Affectional responses in the infant monkey; orphaned baby monkeys develop a strong and persistent attachment to inanimate surrogate mothers. *Science* 130:421-432.1959).

Hart J G, M., and Cicchetti, D. (Salivary cortisol in maltreated children: Evidence of relations between neuroendocrine activity and social competence. . *Development and Psychopathology*, 7:11–26.1995).

Heim C, Ehlert U, Hellhammer DH (The potential role of hypocortisolism in the pathophysiology of stress-related bodily disorders. *Psychoneuroendocrinology* 25:1-35.2000).

Hennessy MB, Kaplan JN (Influence of the maternal surrogate on pituitary-adrenal activity and behavior of infant squirrel monkeys. *Dev. Psychobiol.* 15:423-431.1982).

Higley J, Hasert M, Suomi S, Linnoila M (The serotonin reuptake inhibitor sertraline reduces excessive alcohol consumption in nonhuman primates: effect of stress. *Neuropsychopharmacology* 18:431-443.1998).

Higley JD, Mehlman PT, Poland RE, Taub DM, Vickers J, Suomi SJ, Linnoila M (CSF testosterone and 5-HIAA correlate with different types of aggressive behaviors. *Biol. Psychiatry,* 40:1067-1082.1996).

Higley JD, Suomi SJ, Linnoila M (A longitudinal assessment of CSF monoamine metabolite and plasma cortisol concentrations in young rhesus monkeys. *Biol. Psychiatry,* 32:127-145.1992).

Hinde RA, McGinnis L (Some factors influencing the effects of temporary mother-infant separation: some experiments with rhesus monkeys. *Psychol. Med.* 7:197-212.1977).

Hinde RA, Spencer-Booth Y (The behaviour of socially living rhesus monkeys in their first two and a half years. *Anim. Behav.* 15:169-196.1967).

Hinde RA, Spencer-Booth Y, Bruce M (Effects of 6-day maternal deprivation on rhesus monkey infants. *Nature* 210:1021-1023.1966).

Hotchkiss CE, Paule MG (Effect of pair-housing on operant behavior task performance by rhesus monkeys. *Contemp. Top. Lab. Anim. Sci.* 42:38-41.2003).

Huttenlocher PR, Dabholkar AS (Regional differences in synaptogenesis in human cerebral cortex. *J. Comp. Neurol.* 387:167-178.1997).

Kaplan J (Growth and behavior of surrogate-reared squirrel monkeys. *Dev. Psychobiol.* 7:7-13.1974).

Kaufman D, Banerji MA, Shorman I, Smith EL, Coplan JD, Rosenblum LA, Kral JG (Early-life stress and the development of obesity and insulin resistance in juvenile bonnet macaques. *Diabetes* 56:1382-1386.2007).

Kaufman IC, Rosenblum LA (The reaction to separation in infant monkeys: anaclitic depression and conservation-withdrawal. *Psychosom. Med.* 29:648-675.1967).

Kaufman J, Birmaher B, Perel J, Dahl RE, Moreci P, Nelson B, Wells W, Ryan ND (The corticotropin-releasing hormone challenge in depressed abused, depressed nonabused, and normal control children. *Biol. Psychiatry,* 42:669-679.1997).

Kessler RC (Epidemiology of women and depression. *J. Affect Disord.* 74:5-13.2003).

Kraemer GW, Ebert MH, Schmidt DE, McKinney WT (A longitudinal study of the effect of different social rearing conditions on cerebrospinal fluid norepinephrine and biogenic amine metabolites in rhesus monkeys. *Neuropsychopharmacology* 2:175-189.1989).

Kraemer GW, McKinney WT (Interactions of pharmacological agents which alter biogenic amine metabolism and depression--an analysis of contributing factors within a primate model of depression. *J. Affect Disord.* 1:33-54.1979).

Kreppner JM, Rutter M, Beckett C, Castle J, Colvert E, Groothues C, Hawkins A, O'Connor TG, Stevens S, Sonuga-Barke EJ (Normality and impairment following profound early institutional deprivation: a longitudinal follow-up into early adolescence. *Dev. Psychol.* 43:931-946.2007).

Law AJ, Pei Q, Walker M, Gordon-Andrews H, Weickert CS, Feldon J, Pryce CR, Harrison PJ (Early parental deprivation in the marmoset monkey produces long-term changes in hippocampal expression of genes involved in synaptic plasticity and implicated in mood disorder. *Neuropsychopharmacology* 34:1381-1394.2009).

Levine S (Influence of psychological variables on the activity of the hypothalamic-pituitary-adrenal axis. *Eur. J. Pharmacol.* 405:149-160.2000).

Levine S, Mody T (The long-term psychobiological consequences of intermittent postnatal separation in the squirrel monkey. *Neurosci. Biobehav. Rev.* 27:83-89.2003).

Lewis TL, Maurer D (Multiple sensitive periods in human visual development: evidence from visually deprived children. *Dev. Psychobiol.* 46:163-183.2005).

Loman MM, Gunnar MR (Early experience and the development of stress reactivity and regulation in children. *Neurosci. Biobehav. Rev.* 34:867-876.2010).

Lovejoy CO (The origin of man. *Science* 211:341-350.1981).

Lutz C, Well A, Novak M (Stereotypic and self-injurious behavior in rhesus macaques: a survey and retrospective analysis of environment and early experience. *Am. J. Primatol.* 60:1-15.2003).

Lutz CK, Davis EB, Ruggiero AM, Suomi SJ (Early predictors of self-biting in socially-housed rhesus macaques (Macaca mulatta). *Am. J. Primatol.* 69:584-590.2007).

Lyons DM, Yang C, Mobley BW, Nickerson JT, Schatzberg AF (Early environmental regulation of glucocorticoid feedback sensitivity in young adult monkeys. *J. Neuroendocrinol.* 12:723-728.2000).

Maestripieri D (Early experience affects the intergenerational transmission of infant abuse in rhesus monkeys. *Proc. Natl. Acad. Sci. U S A,* 102:9726-9729.2005).

Maestripieri D, McCormack K, Lindell SG, Higley JD, Sanchez MM (Influence of parenting style on the offspring's behaviour and CSF monoamine metabolite levels in crossfostered and noncrossfostered female rhesus macaques. *Behav. Brain Res.* 175:90-95.2006).

Majer M, Nater UM, Lin JM, Capuron L, Reeves WC (Association of childhood trauma with cognitive function in healthy adults: a pilot study. *BMC Neurol.* 10:61.2010).

Marais L, Daniels W, Brand L, Viljoen F, Hugo C, Stein DJ (Psychopharmacology of maternal separation anxiety in vervet monkeys. *Metab. Brain Dis.* 21:201-210.2006).

Martin LJ, Spicer DM, Lewis MH, Gluck JP, Cork LC (Social deprivation of infant rhesus monkeys alters the chemoarchitecture of the brain: I. Subcortical regions. *J. Neurosci.* 11:3344-3358.1991).

Mason WA, Berkson G (Effects of maternal mobility on the development of rocking and other behaviors in rhesus monkeys: a study with artificial mothers. *Dev. Psychobiol.* 8:197-211.1975).

Mason WA, Sponholz RR (Behavior of Rhesus Monkeys Raised in Isolation. *J. Psychiatr. Res.* 81:299-306.1963).

Meyer JS, Novak MA, Bowman RE, Harlow HF (Behavioral and hormonal effects of attachment object separation in surrogate-peer-reared and mother-reared infant rhesus monkeys. *Dev. Psychobiol.* 8:425-435.1975).

Miller R, Mirsky I, Caul W, Sakata T (Hyperphagia and polydipsia in socially isolated rhesus monkeys. *Science,* 165:1027-1028.1969).

Mitchell GD (Persistent behavior pathology in rhesus monkeys following early social isolation. Folia Primatol (Basel) 8:132-147.1968).

Nelson CA, 3rd, Zeanah CH, Fox NA, Marshall PJ, Smyke AT, Guthrie D (Cognitive recovery in socially deprived young children: the Bucharest Early Intervention Project. *Science* 318:1937-1940.2007).

Nelson EE, Winslow JT (Non-human primates: model animals for developmental psychopathology. *Neuropsychopharmacology* 34:90-105.2009).

Novak MF, Sackett GP (Pair-rearing infant monkeys (Macaca nemestrina) using a "rotating-peer" strategy. *Am. J. Primatol.* 41:141-149.1997).

Parker KJ, Maestripieri D (Identifying key features of early stressful experiences that produce stress vulnerability and resilience in primates. *Neurosci. Biobehav. Rev.*2010).

Pollak SD, Nelson CA, Schlaak MF, Roeber BJ, Wewerka SS, Wiik KL, Frenn KA, Loman MM, Gunnar MR (Neurodevelopmental effects of early deprivation in postinstitutionalized children. *Child Dev*, 81:224-236.2010).

Reite M, Kaufman IC, Pauley JD, Stynes AJ (Depression in infant monkeys: physiological correlates. *Psychosom. Med.* 36:363-367.1974).

Reite M, Short R (Nocturnal sleep in isolation-reared monkeys: evidence for enviromental independence. *Dev. Psychobiol.* 10:555-561.1977).

Reite M, Short RA (Nocturnal sleep in separated monkey infants. *Arch. Gen. Psychiatry,* 35:1247-1253.1978).

Richert KA, Carrion VG, Karchemskiy A, Reiss AL (Regional differences of the prefrontal cortex in pediatric PTSD: an MRI study. *Depress Anxiety,* 23:17-25.2006).

Rommeck I, Gottlieb DH, Strand SC, McCowan B (The effects of four nursery rearing strategies on infant behavioral development in rhesus macaques (Macaca mulatta). *J. Am. Assoc. Lab. Anim. Sci.* 48:395-401.2009).

Rosenblum LA, Andrews MW (Influences of environmental demand on maternal behavior and infant development. *Acta Paediatr. Suppl.* 397:57-63.1994).

Rosenblum LA, Coplan JD, Friedman S, Bassoff T, Gorman JM, Andrews MW (Adverse early experiences affect noradrenergic and serotonergic functioning in adult primates. *Biol. Psychiatry,* 35:221-227.1994).

Rosenblum LA, Forger C, Noland S, Trost RC, Coplan JD (Response of adolescent bonnet macaques to an acute fear stimulus as a function of early rearing conditions. *Dev. Psychobiol.* 39:40-45.2001).

Rosenblum LA, Paully GS (The effects of varying environmental demands on maternal and infant behavior. *Child Dev.* 55:305-314.1984).

Roy MA, Wolf RH, Martin LN, Rangan SR, Allen WP (Social and reproductive behaviors in surrogate-reared squirrel monkeys (Saimiri sciureus). *Lab Anim. Sci.* 28:417-421.1978).

Ruppenthal GC, Walker CG, Sackett GP (Rearing infant monkeys (Macaca nemestrina) in pairs produces deficient social development compared with rearing in single cages. *American Journal of Primatology* 25:103-113.1991).

Sabatini MJ, Ebert P, Lewis DA, Levitt P, Cameron JL, Mirnics K (Amygdala gene expression correlates of social behavior in monkeys experiencing maternal separation. *J. Neurosci.* 27:3295-3304.2007).

Sackett GP (Some persistent effects of different rearing conditions on preadult social behavior of monkeys. *J. Comp. Physiol. Psychol.* 64:363-365.1967).

Sanchez MM (The impact of early adverse care on HPA axis development: nonhuman primate models. *Horm. Behav.* 50:623-631.2006).

Sanchez MM, Hearn EF, Do D, Rilling JK, Herndon JG (Differential rearing affects corpus callosum size and cognitive function of rhesus monkeys. *Brain Res.* 812:38-49.1998).

Sanchez MM, Noble PM, Lyon CK, Plotsky PM, Davis M, Nemeroff CB, Winslow JT (Alterations in diurnal cortisol rhythm and acoustic startle response in nonhuman primates with adverse rearing. *Biol. Psychiatry,* 57:373-381.2005).

Schneider ML, Suomi SJ (Neurobehavioral assessment in rhesus monkey neonates (Macaca mulatta): developmental changes, behavioral stability, and early experience. *Infant Behavior and Development* 15:155-177.1992).

Seay B, Hansen E, Harlow H (MOTHER-INFANT SEPARATION IN MONKEYS*. *Journal of Child Psychology and Psychiatry* 3:123-132.1962).

Shannon C, Champoux M, Suomi SJ (Rearing condition and plasma cortisol in rhesus monkey infants. *Am. J. Primatol.,* 46:311-321.1998).

Shannon C, Schwandt ML, Champoux M, Shoaf SE, Suomi SJ, Linnoila M, Higley JD (Maternal absence and stability of individual differences in CSF 5-HIAA concentrations in rhesus monkey infants. *Am. J. Psychiatry,* 162:1658-1664.2005).

Shansky RM, Morrison JH (Stress-induced dendritic remodeling in the medial prefrontal cortex: effects of circuit, hormones and rest. *Brain Res.,* 1293:108-113.2009).

Siegel SJ, Ginsberg SD, Hof PR, Foote SL, Young WG, Kraemer GW, McKinney WT, Morrison JH (Effects of social deprivation in prepubescent rhesus monkeys: immunohistochemical analysis of the neurofilament protein triplet in the hippocampal formation. *Brain Res.,* 619:299-305.1993).

Spencer-Booth Y, Hinde RA (The effects of 13 days maternal separation on infant rhesus monkeys compared with those of shorter and repeated separations. *Animal Behaviour* 19:595-605.1971).

Spinelli S, Chefer S, Suomi SJ, Higley JD, Barr CS, Stein E (Early-life stress induces long-term morphologic changes in primate brain. *Arch. Gen. Psychiatry,* 66:658-665.2009).

Struble RG, Riesen AH (Changes in cortical dendritic branching subsequent to partial social isolation in stumptailed monkeys. *Dev. Psychobiol.* 11:479-486.1978).

Suomi SJ (Surrogate rehabilitation of monkeys reared in total social isolation. *J. Child Psychol. Psychiatry,* 14:71-77.1973).

Suomi SJ (Maternal Behavior by Socially Incompetent Monkeys: Neglect and Abuse of Offspring. *J. Pediatr. Psychol.* 3:28-34.1978).

Suomi SJ (Early stress and adult emotional reactivity in rhesus monkeys. Ciba Found Symp 156:171-183; discussion 183-178.1991).

Suomi SJ (Early determinants of behaviour: evidence from primate studies. *Br. Med. Bull.* 53:170-184.1997).

Suomi SJ (Mother–infant attachment, peer relationships, and the development of social networks in rhesus monkeys. . *Human Development* 48:67-79.2005).

Suomi SJ, Harlow HF, Kimball SD (Behavioral effects of prolonged partial social isolation in the rhesus monkey. *Psychol. Rep.* 29:1171-1177.1971).

Suomi SJ, Harlow HF, Novak MA (Reversal of social deficits produced by isolation rearing in monkeys. *Journal of Human Evolution* 3:527-534.1974).

Suomi SJ, Mineka S, DeLizio RD (Short- and long-term effects of repetitive mother-infant separations on social development in rhesus monkeys. *Developmental Psychology* 19:770-786.1983).

Tottenham N, Sheridan MA (A review of adversity, the amygdala and the hippocampus: a consideration of developmental timing. *Front Hum. Neurosci.* 3:68.2009).

Uylings HB, van Eden CG (Qualitative and quantitative comparison of the prefrontal cortex in rat and in primates, including humans. *Prog. Brain Res.* 85:31-62.1990).

Wallen K, Goldfoot DA, Goy RW (Peer and maternal influences on the expression of foot-clasp mounting by juvenile male rhesus monkeys. *Dev. Psychobiol.* 14:299-309.1981).

Washburn DA, Rumbaugh DM (Impaired performance from brief social isolation of rhesus monkeys (Macaca mulatta): a multiple video-task assessment. *J. Comp. Psychol.* 105:145-151.1991).

Winslow JT, Noble PL, Lyons CK, Sterk SM, Insel TR (Rearing effects on cerebrospinal fluid oxytocin concentration and social buffering in rhesus monkeys. *Neuropsychopharmacology* 28:910-918.2003).

Worlein JM, Sackett GP (Social development in nursery-reared pigtailed macaques (Macaca nemestrina). *Am. J. Primatol.* 41:23-35.1997).

Wright P (Patterns of paternal care in primates. *International Journal of Primatology* 11:89-102.1990).

Zhang TY, Meaney MJ (Epigenetics and the environmental regulation of the genome and its function. *Annu. Rev. Psychol.* 61:439-466, C431-433.2010).

In: Monkeys: Biology, Behavior and Disorders
Editor: Rachel M. Williams, pp 77-95
ISBN: 978-1-61209-911-8

Chapter 3

PARENT-INFANT RELATIONSHIP IN MARMOSETS

***Atsuko Saito*[1] *and Katsuki Nakamura*[2,*]**

[1]Department of Cognitive and Behavioral Science, Graduate School of Arts and Sciences, the University of Tokyo, 3-8-1 Komaba, Meguro-ku, Tokyo 153-8902, Japan

[2]Section of Cognitive Neuroscience, Department of Behavioral and Brain Sciences, Primate Research Institute, Kyoto University, 41-2 Kanrin, Inuyama, Aichi 484-8506, Japan

ABSTRACT

Although paternal behavior is rare in mammals including primates, males take an active role in rearing infants among callitrichid species. The repertoire of male–infant interactions observed in these species is similar to that observed in humans, such as carrying, protecting, food sharing, grooming, playing, and proximity. Callitrichid species could be a suitable model for research on paternal behavior as well as maternal behavior and the relationship between parents and infants. In this review, we first survey developmental changes in the relationship between infants and parents. Next, we introduce experiments in which attachment behavior from infants to parents and from parents to infants was investigated and examine whether the behavior differs between fathers and mothers. In addition, we refer to a unique behavior of callitrichid species, namely, food transfer from parents to offspring. Finally, we examine how behavior can be controlled via endocrinological variables.

INTRODUCTION

Parental behavior is rare in mammals (Woodroffe and Vincent 1994), even in primates. However, adult males of some primate species provide care of infants and juveniles (van Schaik and Paul 1996). Among such primate species, in the common marmoset (*Callithrix jacchus*) and other callitrichid species (marmosets and tamarins), breeding males and older siblings take an active role in addition to breeding females. Their behavior has been observed

* E-mail: katsuki@pri.kyoto-u.ac.jp, Tel and Fax: +81-568-63-0558

not only in captive groups (Ingram 1977; Box 1977; Locke-Haydon and Chalmers 1983; Locke-Haydon 1984; Tardif et al. 1984; Tardif et al. 1986; Rothe et al. 1993; Ximenes and Sousa 1996; Tardif et al. 1998; Mills et al. 2004; Yamamoto and Box 1997; Yamamoto et al. 2008), but also in wild groups (Yamamoto et al. 2008).

A repertoire of male–infant interactions observed in these species is similar to such interactions in humans, such as carrying, protecting, food sharing, grooming, playing, and proximity (Whiten 1987). Common marmosets are small (250-450g body weight in captivity) and eat exudates (gums, resins, saps) by gouging holes in trees, nectar, fruits, insects, etc. They occupy small home ranges (0.5-6.5 ha) in the Amazon basin of the northeastern part of Brazil (Rylands and de Faria 1993). Group size is reported in the wild as 3-15 members (Ferrari and Ferrari 1989).

The social structure varies from one male-one female to multimale-multifemale groups, but reports of polyandrous groups are rare (Digby et al. 2007). Females sexually mature in about 1 year, and males sexually mature in about 1.5 years (Abbott and Hearn 1978; Harvey et al. 1987). Typically, one pair of a dominant female and male in each group monopolize the reproduction (Saltzman 2003). The breeding female produces litters of 2-3 infants (birth weight approximately 30 g (Tardif et al. 1998)) at roughly 6-month intervals (Tardif et al. 2003). After 1 to 2 weeks of the parturition, the females' estrus comes and they ovulate (Kholkute 1984; Dixson and Lunn 1987); if they get pregnant, they are raising their fetuses while carrying infants. There are high costs for infant carrying in callitrichid species because of twining: a high maternal-infant weight rate (Leutenegger 1973), metabolic expenses (Sanchez et al. 1999; Achenbach and Snowdon 2002), and reduced mobility (Schradin and Anzenberger 2001). The energy and ecological demands of rearing 2 "heavy" infants has been suggested as the cause for the existence of a cooperative breeding system. These traits of common marmosets could make them suitable animal models to understand mechanisms of not only maternal but also paternal behavior and their biological bases.

Parental Behavior in Common Marmosets

Developmental Changes in the Parent-Infant Relationship

Infants of common marmosets have the ability to cling to their caretakers by grabbing their fur. Typically, they are carried on the backs of caretakers (Figure 1). During the first week, mothers or fathers are the main carriers of infants. After the second or third week, older siblings take part in carrying the infants (Ingram 1977; Miller and Lago 1990). One characteristic of common marmosets is that they show extensive paternal care immediately after birth, which is not the case in other callitrichid species (Yamamoto 1993).

After the third week, infants gradually start to behave independently, and the time spent by caretakers carrying infants is then reduced. Beginning around the fourth week, infants have an increased interest in their physical environment and in foods that others are eating. It is at this time that they begin to taste solid food.

At the age of 5 weeks, most infants are independent during half of the observation time. At the age of 10 weeks, they are seldom observed on caretakers' backs (Ingram 1977; Miller and Lago 1990).

Figure 1. Infant marmosets are typically carried on the backs of their caretakers. In this photograph, an older sibling is carrying two infants. Arrows indicate the heads of infants.

However, even after the age of 10 weeks, in fearful situations infants rush to caretakers and climb on their backs (personal observation). Weaning occurs from weeks 8-15 (Yamamoto 1993). Independent feeding is promoted by food sharing from caretakers. The details of food sharing are described later. In many cases, a caretaker smoothly transfers an infant to another caretaker. Usually, the recipient caretaker takes the infant from the back of the other caretaker. Non-mother members of groups are highly motivated to carry infants (Schradin and Anzenberger 2003; Zahed et al. 2008). In social units, there can be competition among older siblings for carrying infants (Yamamoto and Box 1997). However, sometimes caretakers reject infants by biting or rubbing them (Figure 2), and then the infants are left alone for a few minutes. This could happen even when the infants are 1 or 2 weeks old in captive groups (personal observation). These rejected infants emit distress calls that usually stimulate the adult animals to perform the nursing response (Epple 1968). Transfers of infants between caretakers are controlled not only by caretakers but also by infants themselves. They spontaneously transfer from one caretaker to another (Tardif et al. 2002). Therefore, infants passively and actively have contact with members other than their mothers relatively soon after birth. The division of caretaking behavior is affected by the presence or absence of alloparental resources such as older siblings. As group sizes become larger, the time spent carrying infants by each animal decreases (Yamamoto and Box 1997). The presence of older siblings reduces the parents' costs of carrying infants. However, the mothers' carrying of infants at a certain frequency is indispensable to prevent lactation reduction. That is, the presence of older siblings as helpers affects the fathers' participation in carrying infants more than the mothers' participation (Rothe et al. 1993; Ximenes and Sousa 1996). Licking the anogenital regions of infants, a caretaking behavior that stimulates urination, is mainly conducted by a mother, even in a group where a father and older siblings exist (Kaplan and Rogers 1999).

Figure 2. Caretakers sometimes reject infants by biting or rubbing them and removing them. In this photograph, a mother is rejecting her infant by biting, and the infant is emitting distress calls.

This division of caretaking behavior differs not only in quantity, but also in quality among various caretakers. Infants have different relationships with different caretakers, and each relationship changes according to development. For example during the 6th week after birth, carrying episodes on mothers are terminated more frequently by mothers than by infants, while those on fathers are terminated more often by infants than by fathers (Locke-Haydon and Chalmers 1983). In fearful situations, infants seek protection from their fathers, who are most involved in carrying infants in cotton-top tamarins (Kostan and Snowdon 2002). In common marmosets, however, removal of the father does not distress infants as long as their mothers remain available to them (Arruda et al. 1986). In callitrichid species, it is suggested that infants can form multiple, different relationships with their various caretakers (Maestripieri 2003).

Individual Recognition and Selective Approach in the Parent-Infant Relationship

In many species in which the parents take care of their offspring, infants form an emotional bond of attachment selectively with their mothers (Bowlby 1969). This bond functions in distinguishing the mother from others, maintaining contact with them, and recovering this contact if it is destroyed. In primates, the development of attachment has been extensively studied in macaques. Mothers seem to be able to distinguish their infants at the age of one week (Jensen 1965; Jovanovic et al. 2000). While macaque infants could distinguish their mothers from other females visually within a few days (Yamaguchi 2000) and tactually or auditorily in a few weeks (Masataka 1985; Negayama and Honjo 1986) after birth, approach behavior selective for the mother appears within a couple of months after birth (Nakamichi and Yoshida 1986). Thus, one factor of attachment in infants, their ability to recognize their caretaker, could develop early in their life, while approach behavior selective for their mother is observed relatively later in macaques probably because it depends on the development of their locomotor ability. As described above, while macaque infants form attachments selectively for their mothers, infant marmosets may form attachments to their various caretakers. Thus, there is a possibility that their attachments differ from those of macaques. Here, we will describe studies investigating the attachment of infant marmosets.

On the one hand, parent marmosets do not seem to distinguish their own infants from other infants at early stages of development. When presented with infants of less than 2 weeks old, mothers approach quickly and attempt to carry the unknown infants as well as their own (Saltzman and Abbott 2005). However, Hilario and Ferrari (2010) reported that a mother with few-day-old infants killed the infants of other subordinate females. Fathers of 2-4-week-old infants also do not distinguish their own infants from others (Zahed et al. 2008); they approach and carry both equally. It is still unclear when parents start to distinguish their infants from others. On the other hand, infant marmosets could recognize their mothers or fathers within a few days or a few weeks after birth. Infants selectively climb onto their parents (Tardif et al. 2002), and they often rejected an unfamiliar male trying to carry them (Zahed et al. 2008). We tested selective approach by infant marmosets for their parents in a social preference test, in which a parent and a non-parent adult were presented simultaneously to the infant. The selective approach could be observed by 10 weeks of age (Saito et al. 2010) (Figure 3). This approach behavior for either the father or the mother does not differ. Unlike macaques, attachments are similarly formed for both mothers and fathers with infant marmosets. Unlike macaque mothers, in common marmosets both mothers and fathers do not recognize their own infants, at least during the first few weeks after birth. On the other hand, similar to macaques, the infants' ability to distinguish their parents from others develops early in life, while the selective approach behavior appears relatively late in marmosets. This developmental order is the same as macaques. There is no difference in attachment between fathers and mothers in the social preference test. In marmosets, basically all group members are caretakers for infants. One exception is dominant females in polygynous groups, which commit infanticide (Bezerra et al. 2007; Hilario and Ferrari 2010). Considering these social environments, the early development of the infants' ability to distinguish their group members from others, not only their parents but also their older siblings, may increase their survival rate. The attachment between infants and older siblings remains unknown.

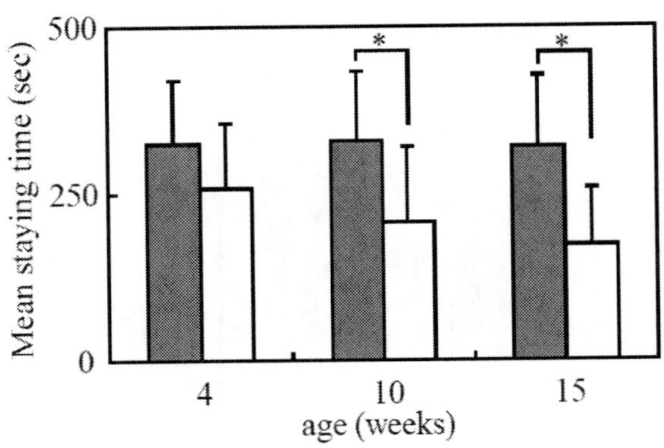

Figure 3. The mean time that infants stayed near parents (filled bars) and near non-parents (open bars) is indicated as determined in our social preference test. The data from both father and mother trials were summed. Asterisks indicate that infants stayed significantly longer near parents than near non-parents (Modified from Saito A, *et al*. Development of infant common marmosets' (*Callithrix jacchus*) preference for their parents over adults from another group. Primates, in press.).

Food Transfer from Caretakers to Offspring

Among nonhuman primates in general, food transfer from adults to their offspring is relatively rare. However, it is reported commonly in apes as well as callitrichids, including common marmosets (*Callithrix jacchus*) (Brown et al. 2004). This peculiar behavior may result from their cooperative breeding system. This behavior is not limited to the parent-infant relationship; it is also frequently observed among adults. In captivity, animals will take pieces of preferred food out of another's oral cavity directly (Kasper et al. 2008). Even in experimental situations, adult marmosets share food (Werdenich and Huber 2002).

Food transfer from parents to offspring is observed during and after weaning (Brown et al. 2005; Saito et al. 2008). Different hypotheses for the functions of food transfer from parents to offspring have been reviewed by Brown *et al.* (2004). The informational hypothesis argues that food transfer plays a role in the acquisition of knowledge about diet choices and food processing skills. Another is the nutritional hypothesis, which suggests that food transfer is considered to provide nutrients to infants during weaning when they are susceptible to food shortage or, more generally, while they develop as independent foragers. The informational hypothesis predicts that infants show higher interest in, beg for, and receive novel food items more frequently than familiar, common food items. This is supported by data from free-ranging golden lion tamarins (*Leontopithecus rosalia*) (Rapaport 2006) and by experiments with captive callitrichid species (common marmosets (Brown et al. 2005) and golden lion tamarins (Rapaport 1999)). Based on this hypothesis, it is also predicted that adults will more willingly transfer novel, uncommon food items to infants than familiar, common food items. Some studies obtained conflicting findings regarding the behavior of adults. Adult animals will refuse to transfer food more frequently when it is novel than when it is familiar (Feistner and Chamove 1986; Price and Feistner 1993; Rapaport 1999; Brown et al. 2005). In contrast, based on the nutritional hypothesis, it is predicted that begging and transfer of food are most frequently observed around the time of weaning. In longitudinal studies, the frequency of food transfer to infants increased to a maximum around the weaning period and then decreased gradually in cotton-top tamarins (*Saguinus oedipus*) (Feistner and Price 1990), black lion tamarins (*Leontopithecus chrysopygus*) (Feistner and Price 2000), and pied bare-faced tamarins (*Saguinus bicolor bicolor*) (Price and Feistner 2001).

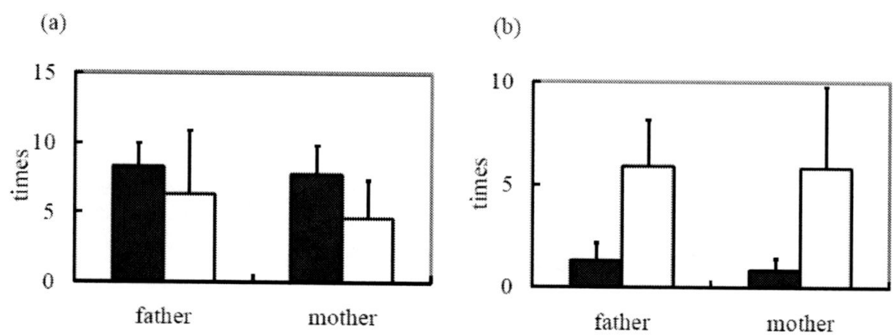

Figure 4. Both fathers and mothers rejected older offspring (29-49 weeks old, filled bars) more frequently than younger offspring (7-15 weeks old, open bars) and (a) transferred food more often to younger offspring than to older offspring (b) when foods were presented in a one-parent-one offspring pair condition (Drawn from Saito A, *et al.* Food transfer in common marmosets: parents change their tolerance depending on the age of offspring. American Journal of Primatology, 70: 999-1002_2008.).

Although not necessarily exclusive, there are two possible explanations for this phenomenon: (1) the begging behavior of infants changes with age; and (2) the tolerance of adults to such begging changes depending on the age of their infants. The rate of successful begging – the ratio of the amount of begging made by infants resulting in food transfer to the total amount of begging – has been shown to remain the same for infants with ages ranging from 4 or 5 to 26 weeks (Feistner and Price 2000; Price and Feistner 2001). This suggests that a parent's positive response to begging for food by its infant remains at least through this period.

Although in these studies foods were presented to a whole family, it was difficult to answer whether the tolerance shown by a particular individual changes according to the infants' age and which of the family members is most tolerant to the infants. Moreover, under this "whole-family food presenting" situation, the behavior of the parents might be affected by distraction from the presence of other family members. To answer these questions, we conducted a study where a food was presented to a one parent-one offspring pair. Consequently, parent marmosets refused older weaned offspring more frequently than younger offspring around the weaning period and transferred food more often to younger offspring than to older offspring (Saito et al. 2008) (Figure 4). These results suggest that both fathers and mothers are more tolerant to weanlings, but their tolerance decreases as offspring mature.

Some researchers have shown that fathers transfer food to offspring more frequently than other family members among cotton-top tamarins (Kostan and Snowdon 2002; Roush and Snowdon 2001) and pied bare-faced tamarins (Price and Feistner 2001). However, in our study there was no difference in all behavioral categories between fathers and mothers in one-on-one experimental situations (Saito et al. 2008). Therefore, it is suggested that this difference was not caused by a difference in tolerance between fathers and mothers. In the natural situation, fathers may more actively transfer food to offspring than mothers in order to assist mothers, who may become pregnant while suckling previous offspring, in allocating resources for the next offspring. Alternatively, offspring might choose their partners by how they can obtain food.

There are two important findings from our study. First, parent marmosets change their behavior to their infants depending on the infant's age. The food transfer behavior of parent marmosets does not differ in quality and quantity between mothers and fathers.

Endocrinological Variables and Parental Behavior

On the one hand, expression of parental behavior is affected by the experience of the parents as a helper (Tardif et al. 1984) and by the presence of helpers (older siblings), as described above. On the other hand, the behavior is also explained by physiological aspects (Pryce 1996). These factors are not independent and are related each other. Endocrinological effects on parental behavior and its neural mechanisms have been studied minutely in sheep (Kendrick et al. 1997) and rats (Numan 2007). The subject of these studies is mainly maternal behavior. In callitrichid species, it is possible to study the physiological aspects of caretaking behavior and the relationship between such caretaking behavior and hormones of not only mothers, but also fathers and older siblings. Many hormones, including steroids (estradiol, progesterone, and testosterone) and peptides (prolactin, vasopressin, and oxytocin), are suggested to play some role in the onset and maintenance of parental behavior in callitrichid species.

Hormonal Changes across Pregnancy and Parturition

In female marmosets, drastic changes in the levels of progesterone, estradiol, prolactin, and oxytocin occur around birth and during lactation. These hormones are also implicated in maternal behavior. Because fathers do not seem to show such changes, it is debatable whether the mechanisms underlying maternal and paternal behavior are the same (Wynne-Edwards 2001; Wynne-Edwards and Timonin 2007; Fernandez-Duque et al. 2009). During the 20-21 weeks of pregnancy, hormonal changes in females are as follows. During the first 10 weeks of gestation, the serum progesterone level is stable. After that, it increases until 1-2 weeks before parturition and decreases suddenly before parturition (Pryce et al. 1993; Moro et al. 1995). Estradiol also increases after the 10th week of pregnancy and decreases after parturitions (Pryce et al. 1993; Saltzman and Abbott 2005). The prolactin level rises gradually after the 10th week of gestation. After parturitions, the level is higher than that during pregnancy (McNeilly et al. 1981; Moro et al. 1995; Torii et al. 1998; Saltzman and Abbott 2005).

Not only mothers, but also fathers in callitrichid species show hormonal changes during pregnancy of its mate. The prolactin level in father marmosets increases in the latter half of the gestation period (Torii et al. 1998) and in cotton-top tamarins at the end of pregnancy for "experienced" fathers that have experienced the birth of their infants (Ziegler and Snowdon 2000; Ziegler et al. 2004b; Almond et al. 2008). In common marmosets, however, there is a discrepancy. A study has reported that male marmosets do not show an increase in prolactin secretion shortly before birth of their infants (Schradin and Anzenberger 2004). In contrast, another study has reported that after the birth of their infants, prolactin levels rise more than twofold in father marmosets (Dixson and George 1982; Schradin and Anzenberger 2004). During the postpartum period, prolactin levels are reportedly higher in experienced fathers in comparison to other males in cotton-top tamarins (Ziegler et al. 1996).

The estradiol level also increases in experienced father cotton-top tamarins at the end of the pregnancy of its mate (Ziegler et al. 2004b). In black tufted-ear marmosets, the estradiol level in the 4 weeks before delivery is higher than that in the following 4 weeks (Nunes et al. 2000). Patterns of changes in the testosterone level in fathers according to the pregnancy and delivery of their mates seem to differ among species. The testosterone level increases at the end of their mate's pregnancy in father cotton-top tamarins regardless of their experience (Ziegler and Snowdon 2000; Ziegler et al. 2004b). In cotton-top tamarins, the testosterone level increases more after birth (Ziegler and Snowdon 2000). In black tufted-ear marmosets, the testosterone level is lower postpartum than the prepartum level (Nunes et al. 2001). In common marmosets, no change in testosterone has been associated with fatherhood (Dixson and George 1982). Males also react to the mate's pregnancy by gaining weight (Ziegler et al. 2006). They might be preparing for the energetic cost of fatherhood.

Therefore, although hormonal changes associated with the pregnancy and the birth of their infants in mothers are clear, those in fathers are variable depending on experience and on the species.

Parenthood and Responsiveness to Infants

Do these changes in hormonal level according to one's own or one's mates' pregnancy and birth trigger parental behavior? Could the difference in hormonal levels explain individual differences in parental behavior? In primates, the reproductive condition affects maternal responsiveness to infants. For example, maternal responsiveness increases across pregnancy in pigtail macaques (Maestripieri and Wallen 1995; Maestripieri and Zehr 1998). Pryce *et al.* (1993) investigated maternal responsiveness using an operant conditioning paradigm. In this paradigm, adult female common marmosets were trained to press a bar to raise a door to gain visual access to a replica of an infant marmoset and to simultaneously turn off infant distress vocalizations. Pryce *et al.* reported that the frequency of bar pressing was maximal just before birth in pregnant female marmosets. These results suggest that prepartum changes in estradiol and progesterone levels increase motivation for maternal behavior. However, Saltzman and Abbott (Saltzman and Abbott 2005) hypothesized that females diminish maternal responsiveness at the late stage of pregnancy because of infanticide by females in the late stages of pregnancy. They showed that while postpartum multiparous females approach both their own and unfamiliar infants quickly, the same females take longer to approach unfamiliar infants and spend less time carrying them during pregnancy. Most females in the late stages of pregnancy never carry infants. This seems to be due to the low prolactin and high progesterone and estrogen levels. The authors explained the difference between these two findings by pointing out the possibility that females wanted to stop the distress call by pressing the bar. Moreover, the difference in parity might cause the difference in the results. It should be mentioned that within reproductive conditions, maternal behaviors were not correlated with hormone levels in the study of Saltzman and Abott (2005). Paternal responsiveness is dependent on fatherhood in common marmosets. Zahed *et al.* (2008) compared the paternal responsiveness to infant stimuli (real infant or infant vocalization) between fathers and "non-experienced" males. Fathers took a shorter time to approach infant stimuli and expressed a greater frequency of infant-directed behavior than did the non-experienced males (Zahed et al. 2008). As mentioned above, in marmosets, infanticide is committed only by females, probably because of female-female reproductive competition (Saltzman 2003). Females kill the infants of other subordinate females not only prepartum, but also postpartum (Hilario and Ferrari 2010). However adult male marmosets have not been reported to commit infanticide and are usually extremely tolerant of, attracted to, and parental toward both related and unrelated infants (Zahed et al. 2008). This is a special trait in common marmosets as opposed to many other mammals in which males kill non-descendant conspecific infants (Van Shaik and Janson 2000). In callitrichid species, fatherhood may explain the responsiveness to infants. However, maternal responsiveness to infants may not be explained simply by motherhood. It is interesting that because of hormonal changes in mothers and fathers, maternal hormonal changes are stable, but paternal hormonal changes are not stable.

Hormonal Levels and Parental Behavior: Steroid Hormones

Although infant survivorship may be one of the strong indexes of the quality of parental behavior, there is little evidence that hormonal levels directly relate to the quality of parental behavior in both females and males. In red-bellied tamarins (*Saguinus labiatus*), survivorship of the infants of mothers showing a stable prepartum estradiol level is higher than that of mothers whose prepartum estradiol level decreased significantly (Pryce et al. 1988). In contrast to this finding, females exhibited higher prepartum estradiol levels when their infants did not survive for more than 2 weeks, compared to when their infants did survive for more than 2 weeks in black tufted-ear marmosets (Fite and French 2000). In this research, neither the relationship between progesterone metabolites (pregnanediol glucuronide) and infant survivorship nor the relationship between the ratio of these two hormones and infant survivorship was observed. In black tufted-ear marmosets, there is no difference in testosterone and estradiol levels between fathers whose infants survived to weaning and fathers whose infants died at birth (Nunes et al. 2000).

Some studies have reported a significant relationship between parental behavior and hormonal levels. In black tufted-ear marmosets, maternal carrying effort was negatively correlated with prepartum estradiol levels (Fite and French 2000). Estradiol and testosterone levels are negatively correlated with carrying effort in mother marmosets. The elevation of these hormones is related to the next pregnancy. In black tufted-ear marmosets, if females conceived their offspring while their infants are dependent, they reduce carrying effort and increase these hormonal levels (Fite et al. 2005b; Fite et al. 2005a). In black tufted-ear marmosets, the testosterone level is significantly higher in males that rarely carry infants compared to males that frequently carry infants. The same tendency is observed for estradiol in this species (Nunes et al. 2001). However, in cotton-top tamarins, the levels of testosterone and estradiol are not correlated with carrying time in fathers (Ziegler et al. 2004a).

So far, the number of species investigated is limited, and there is a discrepancy in results among studies. More studies are needed to elucidate the relationship between parental behavior and hormonal levels in callitrichid species.

Oxytocin and Vasopressin

In addition to these steroid and peptide hormones, two other peptide hormones secreted from the pituitary gland are thought to be associated with parental behavior: oxytocin and vasopressin. Oxytocin was originally known as a hormone that stimulates parturition and milk letdown and is implicated in maternal behavior. Infusion of oxytocin into the ventricle initiates maternal behaviors, such as nest building and licking and retrieving pups, in virgin female rats (Pedersen and Prange 1979; Pedersen et al. 1982). In humans, Feldman *et al.* (2007) have shown a correlation between maternal behavior and peripheral oxytocin level. The peripheral oxytocin level in the perinatal period was positively related to maternal bonding behaviors, including gazing at, vocalization for, positive affect toward, and the affectionate touching of infants. There is only one study reporting the correlation between maternal behavior and oxytocin in nonhuman primates. Lactating female rhesus macaques have a higher oxytocin plasma concentration than non-lactating females (Maestripieri et al. 2009).

Vasopressin is a hormone that controls water balance. It has been implicated in social behaviors including social recognition, pair-bonding, and paternal behavior (Wang et al. 1998). The infusion of vasopressin into the lateral septum promotes paternal behavior (grooming, crouching, retrieving, and contact) in prairie voles (Wang et al. 1994). In humans, fathers' vasopressin levels are negatively correlated with their youngest child's age (Gray et al. 2007). Father marmosets show an abundance of vasopressin V1a receptors in their brains compared with individuals that are not fathers (Kozorovitskiy et al. 2006). It is suggested that there is sexual dimorphism of the effects of oxytocin and vasopressin from rodents to primates (Carter 2007). However, some studies have shown the positive relationship between oxytocin and paternal behavior not only in rodents (Gubernick et al. 1995; Parker et al. 2001; Bales et al. 2004), but also in humans (Feldman et al. 2010). Therefore, there is a possibility that oxytocin affects both maternal and paternal behavior in marmosets.

The Causal Effect of Hormonal Levels

Above-mentioned studies investigated the correlation between hormonal levels and parental behavior or responsiveness. However, most of them did not directly demonstrate the causality of parental behavior. For example, physical contact with infants, such as carrying, promotes the production of prolactin in fathers and non-parent males and females in common marmosets (Roberts et al. 2001b; Mota et al. 2006). Elevation of the prolactin level may be the result of parenting behavior rather than the cause of it. The number of studies challenging the causality of parental behavior is still small in primates.

Pryce *et al.* (1993) successfully made nulliparous, reproductively suppressed females vigorously press a bar for infant stimuli using an exogenous estradiol and progesterone treatment regimen that mimicked the endocrine milieu of late pregnancy. Administration of a dopamine agonist, Bromocriptine, which reduces prolactin secretion, disrupts responsiveness to infants in parentally inexperienced young common marmosets (Roberts et al. 2001a). However, another study using the dopamine agonist, Cabergoline, showed no changes in the paternal behavior of experienced fathers with its family members (Almond et al. 2006). To clarify this contradiction, Ziegler *et al.* (2009) investigated the causality of prolactin in paternal behavior in common marmosets using three conditions: normal pregnancy, decreased prolactin (Cabergoline treatment), and elevated prolactin (human recombinant prolactin treatment) (Ziegler et al. 2009). Although prolactin treatment changed the estradiol levels and body weight in fathers, in their results both elevated and decreased treatments diminished paternal responsiveness to infants. Therefore, the causal relationship of prolactin on paternal behavior has yet to be directly shown.

Research investigating the cause and effect of oxytocin is rare. A nulliparous rhesus macaque that was peripherally administered an oxytocin antagonist, L368,899®, had a reduced interest in infants (Boccia et al. 2007). Intracerebroventricular injection of oxytocin increased touching and lip-smacking to infants of two nulliparous rhesus macaques (Holman and Goy 1995). As the number of subjects in these studies was very small, it is difficult for us to form a conclusion. In callitrichid species, so far no research has investigated the effect of oxytocin on maternal behavior. We conducted a study and demonstrated that five father marmosets increased their tolerance to offspring after an intracerebroventricular injection of oxytocin (Saito and Nakamura 2008). Therefore, there is a possibility that paternal behavior is mediated by oxytocin in common marmosets.

There is no doubt that endocrinological variables are associated with parenthood and that some endocrinological variables influence parental behavior. However, some evidence shows that the relationship between individual differences in behavior and hormone levels is not significant. Moreover, causalities of many hormones have not been well proven. Changes in parental behavior across pregnancy and into the postpartum period may be facilitated by changes in hormone levels or brain structures, but individual differences in behavior may be more closely linked to nonhormonal factors, such as temperament or experience (Maestripieri 1999).

CONCLUSIOIN

Common marmosets have a unique breeding system in which fathers and older siblings take an important role in parenting. Because of such a breeding system, infant common marmosets seem to make various bonds with family members, while infants of other non-human primates built one bond with their mother (or main caretaker). They also have characteristic behavioral repertoires: food transfer from parents to offspring, imitation (Voelkl and Huber 2000, 2007), sharing of food by dominant individuals with subordinates in the group (even those without kin relationships) (Kasper et al. 2008), and unsolicited prosocial behavior (Burkart et al. 2007). These behavioral repertoires may be facilitated by the cooperative breeding system of marmosets (Burkart and van Shaik 2010). Needless to say, humans are also cooperative breeders. The physiological aspects of maternal and paternal behavior are currently being examined. For example, observed paternal and maternal behavior has a positive correlation with oxytocin level after interaction with infants (Feldman et al. 2010). However, the relationships do not directly indicate the causal effects of hormone level on parental behavior, as shown in the marmoset prolactin study. More precise experiments on animals are needed to elucidate the effects on various hormones on parental behavior. The common marmoset can be a suitable model for this purpose.

ACKNOWLEDGMENTS

Our research was partly supported by the Japan Society for the Promotion of Science (no. 21700284 to A. Saito), CREST, the Japan Science and Technology Agency, and a Research Grant (20B-10) for Nervous and Mental Disorders from the Ministry of Health, Labour and Welfare, Japan (to K. Nakamura).

REFERENCES

Abbott DH, Hearn JP (1978) Physical, hormonal and behavioural aspects of sexual development in the marmoset monkey, *Callithrix jacchus*. *J. Reprod. Fertil.* 53 (1):155-166.

Achenbach GG, Snowdon C (2002) Costs of caregiving: Weight loss in captive adult male cotton-top tamarins (*Saguinus oedipus*) following the birth of infants. *Int. J. Primatol.* 23 (1):179-189.

Almond RE, Brown GR, Keverne EB (2006) Suppression of prolactin does not reduce infant care by parentally experienced male common marmosets (*Callithrix jacchus*). *Horm. Behav.* 49 (5):673-680.

Almond RE, Ziegler TE, Snowdon CT (2008) Changes in prolactin and glucocorticoid levels in cotton-top tamarin fathers during their mate's pregnancy: The effect of infants and paternal experience. *Am. J. Primatol.* 70 (6):560-565.

Arruda MF, Yamamoto ME, Bueno OFA (1986) Interactions between parents and infants, and infants-father separation in the common marmoset (*Callithrix jacchus*). *Primates* 27 (2):215-228.

Bales KL, Kim AJ, Lewis-Reese AD, Sue Carter C (2004) Both oxytocin and vasopressin may influence alloparental behavior in male prairie voles. *Horm. Behav.* 45 (5):354-361.

Bezerra BM, Da Silva Souto A, Schiel N (2007) Infanticide and cannibalism in a free-ranging plurally breeding group of common marmosets (*Callithrix jacchus*). *Am. J. Primatol.* 69 (8):945-952.

Boccia ML, Goursaud AP, Bachevalier J, Anderson KD, Pedersen CA (2007) Peripherally administered non-peptide oxytocin antagonist, l368,899((r)), accumulates in limbic brain areas: A new pharmacological tool for the study of social motivation in non-human primates. *Horm. Behav.* 52 (3):344-351.

Bowlby J (1969) Attachment and loss, vol. 1 attachment. Harmondsworth, Middlesex.

Box HO (1977) Quantitative data on the carrying of young captive monkeys (*Callithrix jacchus*) by other members of their family groups. *Primates* 18 (2):475-484.

Brown GR, Almond RE, Bates NJ (2005) Adult-infant food transfer in common marmosets: An experimental study. *Am. J. Primatol.* 65 (4):301-312.

Brown GR, Almond REA, van Bergen Y (2004) Begging, stealing and offering: Food transfer in non-human primates. *Advances in the Study of Behaviour* 34:265-295.

Burkart JM, Fehr E, Efferson C, van Schaik CP (2007) Other-regarding preferences in a non-human primate: Common marmosets provision food altruistically. *Proc. Natl. Acad. Sci. U S A* 104 (50):19762-19766.

Burkart JM, van Shaik CP (2010) Cognitive consequences of cooperative breeding in primates? *Animal Cognition* 13 (1):1-19.

Carter CS (2007) Sex differences in oxytocin and vasopressin: Implications for autism spectrum disorders? *Behav. Brain Res.* 176 (1):170-186.

Digby LJ, Ferrari SF, Saltzman W (2007) Callitrichines: The role of competition in cooperatively breeding species. In: Campbell CJ (ed) *Primates in perspective*. Oxford University Press, pp 85-105.

Dixson AF, George L (1982) Prolactin and parental behaviour in a male new world primate. *Nature* 299 (5883):551-553.

Dixson AF, Lunn SF (1987) Post-partum changes in hormones and sexual behaviour in captive groups of marmosets (*Callithrix jacchus*). *Physiol. Behav.* 41 (6):577-583.

Epple G (1968) Comparative studies on vocalization in marmoset monkeys (*hapalidae*). *Folia Primatologica,* 8:1-40.

Feistner AT, Price EC (1990) Food-sharing in cotton-top tamarins (*Saguinus oedipus*). *Folia Primatol.* (Basel) 54 (1-2):34-45.

Feistner AT, Price EC (2000) Food sharing in black lion tamarins (*Leontopithecus chrysopygus*). *Am. J. Primatol.* 52 (1):47-54.

Feistner ATC, Chamove AS (1986) High motivation toward food increases food-sharing in cotton-top tamarins. *Developmental Psychobiology,* 19 (5):439-452.

Feldman R, Gordon I, Schneiderman I, Weisman O, Zagoory-Sharon O (2010) Natural variations in maternal and paternal care are associated with systematic changes in oxytocin following parent-infant contact. *Psychoneuroendocrinology* 35 (8):1133-1141.

Feldman R, Weller A, Zagoory-Sharon O, Levine A (2007) Evidence for a neuroendocrinological foundation of human affiliation: Plasma oxytocin levels across pregnancy and the postpartum period predict mother-infant bonding. *Psychol. Sci.* 18 (11):965-970.

Fernandez-Duque E, Valeggia CR, Mendoza SP (2009) The biology of paternal care in human and nonhuman primates. *Annu. Rev. Anthropol.* 38:115-130.

Ferrari SF, Ferrari MAL (1989) A re-evaluation of the social organisation of the callitrichidae, with reference to the ecological differences between genera. *Folia Primatol.* 52:132-147.

Fite JE, French JA (2000) Pre- and postpartum sex steroids in female marmosets (*Callithrix kuhlii*): Is there a link with infant survivorship and maternal behavior? *Horm. Behav.* 38 (1):1-12.

Fite JE, French JA, Patera KJ, Hopkins EC, Rukstalis M, Ross CN (2005a) Elevated urinary testosterone excretion and decreased maternal caregiving effort in marmosets when conception occurs during the period of infant dependence. *Hormones and Behavior* 47 (1):39-48.

Fite JE, Patera KJ, French JA, Rukstalis M, Hopkins EC, Ross CN (2005b) Opportunistic mothers: Female marmosets (*Callithrix kuhlii*) reduce their investment in offspring when they have to, and when they can. *J. Hum. Evol.* 49 (1):122-142.

Gray PB, Parkin JC, Samms-Vaughan ME (2007) Hormonal correlates of human paternal interactions: A hospital-based investigation in urban jamaica. *Horm. Behav.* 52 (4):499-507.

Gubernick DJ, Winslow JT, Jensen P, Jeanotte L, Bowen J (1995) Oxytocin changes in males over the reproductive cycle in the monogamous, biparental california mouse, peromyscus californicus. *Horm. Behav.* 29 (1):59-73.

Harvey PH, Martin RD, Clutton-Brock TM (1987) Life histories in comparative perspective. In: Smuts BB, Cheney DL, Seyfarth RM, Wrangham RW, Struhsaker TT (eds) *Ptimate societies.* Univ of Chicago Press, Chicago, pp 181-196.

Hilario RR, Ferrari SF (2010) Double infanticide in a free-ranging group of buffy-headed marmosets, *Callithrix flaviceps. Journal of Ethology* 28 (1):195-199.

Holman SD, Goy RW (1995) Experiential and hormonal correlates of care-giving in rhesus macaques. In: Pryce CR, Martin RD, Skuse D (eds) M*otherhood in human and nonhuman primates*. Karger, Basel, pp 87-93.

Ingram JC (1977) Interactions between parents and infants, and the development of independence in the common marmosets (*Callithrix jacchus*). *Anim. Behav.* 25:811-827.

Jensen GD (1965) Mother-infant relationship in the monkey *Macaca nemestrina*: Development of specificity of maternal response to own infant. *J. Comp. Physiol. Psychol.* 59:305-308.

Jovanovic T, Megna NL, Maestripieri D (2000) Early maternal recognition of offspring vocals in rhesus. *Primates* 41:421-428.

Kaplan G, Rogers LJ (1999) Parental care in marmosets (*Callithrix jacchus jacchus*): Development and effect of anogenital licking on exploration. *J. Comp. Psychol.* 113 (3):269-276.

Kasper C, Voelkl B, Huber L (2008) Tolerated mouth-to-mouth food transfers in common marmosets. *Primates* 49 (2):153-156.

Kendrick KM, Da Costa AP, Broad KD, Ohkura S, Guevara R, Levy F, Keverne EB (1997) Neural control of maternal behaviour and olfactory recognition of offspring. *Brain Res. Bull.* 44 (4):383-395.

Kholkute (1984) Plasma estradiol and progesterone levels during post partum ovulation and early pregnancy in the common marmoset *Callithrix jacchus*. *Primates* 25 (4):538-543.

Kostan KM, Snowdon CT (2002) Attachment and social preferences in cooperatively-reared cotton-top tamarins. *Am. J. Primatol.* 57 (3):131-139.

Kozorovitskiy Y, Hughes M, Lee K, Gould E (2006) Fatherhood affects dendritic spines and vasopressin v1a receptors in the primate prefrontal cortex. *Nat. Neurosci.* 9 (9):1094-1095.

Leutenegger CM (1973) Maternal-fetal weight relationships in primates. *Folia Primatol.* 20 (4):280-293

Locke-Haydon J (1984) *Infant-caregiver interactions in common marmosets* (*Callithrix jacchus*): *Consequences of rejections and rub-offs of infants by caregivers Primates* 25 (4):450-461.

Locke-Haydon J, Chalmers NR (1983) The development of infant-caregiver relationships in captive common marmosets (*Callithrix jacchus*) *Int. J. Primatol.* 4 (1):63-81.

Maestripieri D (1999) The biology of human parenting: Insights from nonhuman primates. *Neurosci. Biobehav. Rev.* 23 (3):411-422

Maestripieri D (2003) Attachment. In: Maestripieri D (ed) *Primate psychology*. Harvard University Press, London, pp 108-143.

Maestripieri D, Hoffman CL, Anderson GM, Carter CS, Higley JD (2009) Mother-infant interactions in free-ranging rhesus macaques: Relationships between physiological and behavioral variables. *Physiol. Behav.* 96 (4-5):613-619.

Maestripieri D, Wallen K (1995) Interest in infants varies with reproductive condition in group-living female pigtail macaques (*Macaca nemestrina*). *Physiol. Behav.* 57 (2):353-358.

Maestripieri D, Zehr JL (1998) Maternal responsiveness increases during pregnancy and after estrogen treatment in macaques. *Horm. Behav.* 34 (3):223-230.

Masataka N (1985) Development of vocal recognition of mothers in infant japanese macaques. *Dev. Psychobiol.* 18 (2):107-114.

McNeilly AS, Abbott DH, Lunn SF, Chambers PC, Hearn JP (1981) Plasma prolactin concentrations during the ovarian cycle and lactation and their relationship to return of fertility post partum in the common marmoset (*Callithrix jacchus*). *J. Reprod. Fertil.* 62 (2):353-360.

Miller M, Lago D (1990) Observed pet-owner in-home interactions: Species differences and association with the pet relationship scale. *Anthrozoos* 4 (1):49-54.

Mills DA, Windle CP, Baker HF, Ridley RM (2004) Analysis of infant carrying in large, well-established family groups of captive marmosets (*Callithrix jacchus*). *Primates* 45 (4):259-265.

Moro M, Torii R, Koizumi H, Inada Y, Etoh Y, Miyata H, Tanioka Y (1995) Serum levels of prolactin during the ovarian cycle, pregnancy, and lactation in the common marmoset (*Callithrix-jacchus*). *Primates* 36 (2):249-257.

Mota MTD, Franci CR, de Sousa MB (2006) Hormonal changes related to paternal and alloparental care in common marmosets (*Callithrix jacchus*). *Horm. Behav.* 49 (3):293-302.

Nakamichi M, Yoshida A (1986) Discrimination of mother by infant among japanese macaques (*Macaca fuscata*). *Int. J. Primatol.* 7 (5):481-489.

Negayama K, Honjo S (1986) An experimental study on developmental changes of maternal discrimination of infants in crab-eating monkeys (*Macaca fascicularis*). *Dev. Psychobiol.* 19 (1):49-56.

Numan M (2007) Motivational systems and the neural circuitry of maternal behavior in the rat. *Dev. Psychobiol.* 49 (1):12-21.

Nunes S, Fite JE, French JA (2000) Variation in steroid hormones associated with infant care behaviour and experience in male marmosets (*Callithrix kuhlii*). *Anim Behav* 60 (6):857-865.

Nunes S, Fite JE, Patera KJ, French JA (2001) Interactions among paternal behavior, steroid hormones, and parental experience in male marmosets (*Callithrix kuhlii*). *Horm. Behav.* 39 (1):70-82.

Parker KJ, Kinney LF, Phillips KM, Lee TM (2001) Paternal behavior is associated with central neurohormone receptor binding patterns in meadow voles (*Microtus pennsylvanicus*). *Behav. Neurosci.* 115 (6):1341-1348.

Pedersen CA, Ascher JA, Monroe YL, Prange AJ, Jr. (1982) Oxytocin induces maternal behavior in virgin female rats. *Science* 216 (4546):648-650.

Pedersen CA, Prange AJ, Jr. (1979) Induction of maternal behavior in virgin rats after intracerebroventricular administration of oxytocin. *Proc. Natl. Acad. Sci. U S A* 76 (12):6661-6665.

Price EC, Feistner ATC (1993) Food sharing in lion tamarins: Tests of three hypotheses. *American Journal of Primatology* 31:211-221.

Price EC, Feistner ATC (2001) Food sharing in pied bare-faced tamarins (*Saguinus bicolor bicolor*): Development and individual differences. *International Journal of Primatology*, 22 (2):231-241.

Pryce CR (1996) Socialization, hormones, and the regulation of maternal behavior in nonhuman primates. *Adv. Study Behav.* 25:423-473.

Pryce CR, Abbott DH, Hodges JK, Martin RD (1988) Maternal behavior is related to prepartum urinary estradiol levels in red-bellied tamarin monkeys. *Physiol. Behav.* 44 (6):717-726.

Pryce CR, Dobeli M, Martin RD (1993) Effects of sex steroids on maternal motivation in the common marmoset (*Callithrix jacchus*): Development and application of an operant system with maternal reinforcement. *J. Comp. Psychol.* 107 (1):99-115.

Rapaport LG (1999) Provisioning of young in golden lion tamarins (callitrichidae, *Leontopithecus rosalia*): A test if the information hypothesis. *Etholgy* 105:619-636.

Rapaport LG (2006) Provisioning in wild golden lion tamarins (*Leontopithecus rosalia*): Benefits to omnivorous young. *Behavioral. Ecology,* 17 (2):212-221.

Roberts RL, Jenkins KT, Lawler T, Jr., Wegner FH, Newman JD (2001a) Bromocriptine administration lowers serum prolactin and disrupts parental responsiveness in common marmosets (*Callithrix j. jacchus*). *Horm. Behav.* 39 (2):106-112.

Roberts RL, Jenkins KT, Lawler T, Wegner FH, Norcross JL, Bernhards DE, Newman JD (2001b) Prolactin levels are elevated after infant carrying in parentally inexperienced common marmosets. *Physiol. Behav.* 72 (5):713-720.

Rothe H, Darms K, Koenig A, Radespiel U, Juenemann B (1993) Long-term study of infant-carrying behavior in captive common marmosets (*Callithrix jacchus*): Effect of nonreproductive helpers on the parents' carrying performance, *Int. J. Primatol.* 14 (1):79-93.

Roush RS, Snowdon CT (2001) Food transfer and development of feeding behavior and food-associated vocalizations in cotton-top tamarins. *Ethology* 107 (5):415-429.

Rylands AB, de Faria DS (1993) Habtats, feeding ecology, and home range size in the genus *callithrix*. In: Rylands AB (ed) *Marmosets and tamarins: Systematics, behaviour, and ecology*. Oxford University Press, Oxford, pp 262-272.

Saito A, Izumi A, Nakamura K (2008) Food transfer in common marmosets: Parents change their tolerance depending on the age of offspring. *Am. J. Primatol.* 70 (10):999-1002.

Saito A, Izumi A, Nakamura K (2010) Development of infant common marmosets' (*Callithrix jacchus*) preference for their parents over adults from another group. *Primates*. doi:10.1007/s10329-010-0205-7.

Saito A, Nakamura K Oxytocin makes primate father more tolerant to offspring. In: *Society for Neuroscience 38th Annual meeting*, Washington Convention Center, Washington DC., November, 19 2008 2008.

Saltzman W (2003) Reproductive competition among female common marmosets (*Callithrix jacchus*): Proximate and ultimate causes. In: Jones CB (ed) *Sexual selection and reproductive competition in primates: New perspectives and directions*. American Society of Primatologists, Norman, OK, pp 197-229.

Saltzman W, Abbott DH (2005) Diminished maternal responsiveness during pregnancy in multiparous female common marmosets. *Hormones and Behavior* 47 (2):151-163.

Sanchez S, Pelaez F, Gil-Burmann C, Kaumanns W (1999) Costs of infant-carrying in the cotton-top tamarin (*Saguinus oedipus*). *Am. J. Primatol.* 48 (2):99-111.

Schradin C, Anzenberger G (2001) *Costs of infant carrying in common marmosets, Callithrix jacchus: An experimental analysis Anim Behav* 62 (2):289-295.

Schradin C, Anzenberger G (2003) Mothers, not fathers, determine the delayed onset of male carrying in goeldi's monkey (*Callimico goeldii*). *J. Hum. Evol.* 45 (5):389-399.

Schradin C, Anzenberger G (2004) Development of prolactin levels in marmoset males: From adult son to first-time father. *Horm. Behav.* 46 (5):670-677.

Tardif S, Jaquish C, Layne D, Bales K, Power M, Power R, Oftedal O (1998) Growth variation in common marmoset monkeys (*Callithrix jacchus*) fed a purified diet: Relation to care-giving and weaning behaviors. *Lab. Anim. Sci.* 48 (3):264-269.

Tardif SD, Carson RL, Gangaware BL (1986) Comparison of infant care in family groups of the common marmoset (*Callithrix jacchus*) and the cotton-top tamarin (*Saguinus* oedipus). *American Journal of Primatology* 11 (2):103-110.

Tardif SD, Layne DG, Smucny DA (2002) Can marmoset mothers count to three? Effect of litter size on mother-infant interactions. *Ethology* 108 (9):825-836.

Tardif SD, Richter CB, Carson RL (1984) Effects of sibling-rearing experience on future reproductive success in two species of callitrichidae. *American Journal of Primatology,* 6 (4):377 – 380.

Tardif SD, Smucny DA, Abbott DH, Mansfield K, Schultz-Darken N, Yamamoto ME (2003) Reproduction in captive common marmosets (*Callithrix jacchus*). *Comp. Med.* 53 (4):364-368.

Torii R, Moro M, Abbott DH, Nigi H (1998) Urine collection in the common marmoset (*Callithrix jacchus*) and its applicability to endocrinological studies. *Primates* 39 (4):407-417.

van Schaik CP, Paul A (1996) *Male care in primates: Does it ever reflect paternity? Evolutionary Anthropology: Issue*s, News, and Reviews 5 (5):152-156.

Van Shaik CP, Janson CH (2000) Infanticide by males and its implication. Cambridge University Press, Cambridge.

Voelkl B, Huber L (2000) True imitation in marmosets. *Anim. Behav.* 60 (2):195-202.

Voelkl B, Huber L (2007) Imitation as faithful copying of a novel technique in marmoset monkeys. *PLoS ONE* 2 (7):e611.

Wang Z, Ferris CF, De Vries GJ (1994) Role of septal vasopressin innervation in paternal behavior in prairie voles (*Microtus ochrogaster*). *Proc. Natl. Acad. Sci. U S A* 91 (1):400-404.

Wang Z, Young LJ, De Vries GJ, Insel TR (1998) Voles and vasopressin: A review of molecular, cellular, and behavioral studies of pair bonding and paternal behaviors. *Prog. Brain Res.* 119:483-499.

Werdenich D, Huber L (2002) Social factors determine cooperation in marmosets. *Animal Behaviour,* 64:771-781.

Whiten PL (1987) Infants and adult males. In: Smuts BB, Cheney DL, Seyfarth RM, Wrangham RW, Struhsaker TT (eds) *Primate societies.* The University of Chicago Press, Chicago, pp 343-357.

Woodroffe R, Vincent A (1994) Mothers little helpers - patterns of male care in mammals. *Trends in Ecology and Evolution* 9 (8):294-297.

Wynne-Edwards KE (2001) Hormonal changes in mammalian fathers. *Horm. Behav.* 40 (2):139-145.

Wynne-Edwards KE, Timonin ME (2007) Paternal care in rodents: Weakening support for hormonal regulation of the transition to behavioral fatherhood in rodent animal models of biparental care. *Horm. Behav.* 52 (1):114-121.

Ximenes MFFM, Sousa MBC (1996) *Family composition and the characteristics of parental care during the nursing phase of captive common marmosets* (*Callithrix jacchus*). *Primates* 37 (2):175-183.

Yamaguchi M (2000) Akachan wa kao wo yomu (in Japanese). *Kinokuniyashoten*, Tokyo.

Yamamoto ME (1993) From dependence to sexual maturity: The behavioural ontogeny of callitrichidae. In: Rylands AB (ed) *Marmosets and tamarins: Systematics, behaviour, and ecology.* Oxford University Press, Oxford, pp 235-254.

Yamamoto ME, Albuquerque FS, Lopes NA, Ferreira ES (2008) Differential infant carrying in captive and wild common marmosets (*Callithrix jacchus*). *Acta. Ethologica,* 11 (2):95-99.

Yamamoto ME, Box HO (1997) The role of non-reproductive helpers in infant care in captive *Callithrix jacchus. Ethology* 103 (9):760-771.

Zahed SR, Prudom SL, Snowdon CT, Ziegler TE (2008) Male parenting and response to infant stimuli in the common marmoset (*Callithrix jacchus*). *Am. J. Primatol.* 70 (1):84-92.

Ziegler TE, Jacoris S, Snowdon CT (2004a) Sexual communication between breeding male and female cotton-top tamarins (*Saguinus oedipus*), and its relationship to infant care. *Am. J. Primatol.* 64 (1):57-69.

Ziegler TE, Prudom SL, Schultz-Darken NJ, Kurian AV, Snowdon CT (2006) Pregnancy weight gain: Marmoset and tamarin dads show it too. *Biol. Lett.* 2 (2):181-183.

Ziegler TE, Prudom SL, Zahed SR, Parlow AF, Wegner F (2009) Prolactin's mediative role in male parenting in parentally experienced marmosets (*Callithrix jacchus*). *Horm. Behav.* 56 (4):436-443.

Ziegler TE, Snowdon CT (2000) Preparental hormone levels and parenting experience in male cotton-top tamarins, *Saguinus oedipus*. *Horm. Behav.* 38 (3):159-167.

Ziegler TE, Washabaugh KF, Snowdon CT (2004b) Responsiveness of expectant male cotton-top tamarins, *Saguinus oedipus*, to mate's pregnancy. *Horm. Behav.* 45 (2):84-92.

Ziegler TE, Wegner FH, Snowdon CT (1996) Hormonal responses to parental and nonparental conditions in male cotton-top tamarins, *Saguinus oedipus*, a new world primate. *Horm. Behav.* 30 (3):287-297.

In: Monkeys: Biology, Behavior and Disorders
Editor: Rachel M. Williams, pp 97-117

ISBN: 978-1-61209-911-8

Chapter 4

EXPLORATION AND AMBULATORY BEHAVIOURS IN NORMAL AND FORNIX TRANSECTED MACAQUE MONKEYS IN AN OPEN SPACE

Sze Chai Kwok*

Department of Experimental Psychology, University of Oxford, Oxford, United Kingdom
Neuroimaging Laboratory, Santa Lucia Foundation, Rome, Italy

ABSTRACT

Prompted by the theoretical prediction that damage to the hippocampal system should abolish exploratory behaviour, the present study examined exploratory movements in control monkeys (CON) and monkeys with transection of the fornix (FNX), a major input/output pathway of the hippocampus. CON and FNX monkeys were introduced to a large novel octagonal chamber (approximately 7.4 m^2) for six daily sessions each lasting 20 minutes. Both groups visited, punctuated by stops, the majority of the floor-space of the environment in each of the sessions. The exploratory movements of CON and FNX groups were not significantly different on most of the measures taken over 6 consecutive days. These measures included cumulative distance traveled, number and duration of stops, travelling patterns, and proportion of time spent in each of 12 designated zones of floor-space. The high degree of similarity in behaviour between CON and FNX groups suggests that an intact hippocampal system is not necessary for the display of normal exploratory movement per se. On the other hand, the CON and FNX groups did behave differently on two measures. First, the CON group exhibited a decrement in distance traversed over consecutive epochs within the first test session whereas FNX animals did not. Second, on those days in which the chamber was made visually asymmetrical, the CON animals tended to show a predilection for spending proportionally more time within one particular quadrant of the chamber. These observations are consistent with the idea that interrupting normal hippocampal system function by means of fornix transection is detrimental to learning about the spatial layout of environments. As the first attempt to compare exploratory and ambulatory behaviours

* Corresponding author: Sze Chai Kwok, Address: Neuroimaging Laboratory, Santa Lucia Foundation, Via Ardeatina 306, 00179 Rome (Italy), Telephone: +39 06 5150 1459, Fax:+39 06 5150 1213, Email: sze-chai.kwok@st-hughs.oxon.org

of monkeys with and without fornical damage in a large open space, I argue that while monkeys with fornix transection still display intact locomotor and exploratory behaviour patterns, their new learning of visuospatial context is impeded.

INTRODUCTION

In their seminal book *Hippocampus as a Cognitive Map* (1978), O'Keefe and Nadel postulated a theoretical link between hippocampal circuitry and exploratory behaviour, in which they stated exploration is fundamental in building and updating the internal representation of the spatial layout of an environment and that an intact hippocampus is pivotal to a functional exploratory repertoire. O'Keefe and Nadel's argument that animals with hippocampal lesions would cease to display exploratory behaviour was based upon the hypothesis that the hippocampus serves as a cognitive map requiring environmental information derived from exploration (Clark, Hines, Hamilton, & Whishaw, 2005; O'Keefe & Nadel, 1978). Although the hypothesis that the hippocampus serves as a cognitive map has received extensive examination (e.g. Redish, 1999), there has not been similar focus of research into the predication that the hippocampal system is required for exploration (Clark, Hines, Hamilton, & Whishaw, 2005).

Other workers in the field have advocated similar theories that the hippocampal formation is essential for spatial relational learning and memory (Howard Eichenbaum, 2000; H Eichenbaum, Otto, & Cohen, 1994), and for the formation and use of allocentric representations of space (Lavenex, Amaral, & Lavenex, 2006; Nadel & Hardt, 2004; O'Keefe & Nadel, 1978). Studies have reported that there are alterations in exploratory behaviours displayed by hippocampal lesioned rats in open field tests (Whishaw, Cassal, & Majchrzak, 1994), but other studies have reported that hippocampectomized rats still investigate novel objects and are sensitive to changes in their location (Harley & Martin, 1999). However, until recently there have been no detailed analyses of animals' ongoing movements and movement patterns after hippocampal system damage. For this reason, Clark and colleagues (2005) examined exploratory movements in control rats and rats with hippocampal lesions [produced with the neurotoxin N-methyl D-aspartate (NMDA)] attempting to test the hypothesis that damage to the hippocampus would abolish exploratory behaviour. On four successive days, control and hippocampal rats were placed onto a circular table near to which a large salient visual cue was positioned and their exploratory movements were measured. They found that there was no difference in these measures between control rats and rats with hippocampal lesions on the first test day. However, by the time of the fourth day of testing control animals were found to be less active on most of the measures of exploration, whereas the behaviour of hippocampal rats remained unchanged from that observed on the first day. Thus Clark and colleagues (2005) asserted that the hippocampus may not be necessary for the display of normal exploratory movements by making inference from the high degree of similarity in behaviour between hippocampal lesioned and control rats on Day 1, and the persistence of this behaviour in hippocampal rats on Day 4. This finding also suggests that hippocampectomized rats' behaviour might be related to a spatial memory impairment impeding acquisition of familiarity with the environment that the animal finds itself in (Clark, Hines, Hamilton, & Whishaw, 2005).

A handful of ambulatory studies on nonhuman primates have also been conducted. One of the earliest ambulatory studies on nonhuman primates was carried out by Murray and colleagues (1989 Exp. 1) in which macaque monkeys were trained preoperatively on delayed non-matching to sample in a T-maze, and were then retrained postoperatively to the same criterion they had attained before surgery. The results demonstrate that monkeys with fornix transection are severely impaired on this spatial working memory task requiring locomotion. However, strictly speaking, this task could not be regarded as an exploratory task because behaviour at the choice point in the maze was simply a choice between two spatial directions of movement and so this task is at best qualified as a delayed non-matching to locations test that involves locomotor actions. Interestingly, it has been shown that normal marmoset monkeys can be trained to perform very well at this task too (Easton, Parker, Derrington, & Parker, 2003).

Until more recently, two other studies, both conducted in an open-field, set out to investigate the effects of selective bilateral hippocampal lesions on spatial learning in ambulatory macaque monkeys. The first study was a series of open-field experiments (Hampton, Hampstead, & Murray, 2004) that allowed tethered monkeys to walk about in a large-scale environment. Monkeys with bilateral hippocampal excitotoxic lesions were tested in match-to-location tasks. The results demonstrate that selective hippocampal lesions in monkeys impair memory for location in an open-field. Although the data indicate that the monkeys encoded some allocentric information about the goal locations, the study was not able to demonstrate a reliance on an allocentric representation of space in normal monkeys by ruling out the use of an egocentric strategy; and thus the authors were unable to determine whether their monkeys were using a true 'cognitive map'. In contrast, in Lavenex and colleagues (2006)'s study, freely moving monkeys were trained to forage for food located in six goal locations among 18 locations distributed in an open-field arena. Multiple goals and four pseudorandomly chosen entrance points precluded the monkeys' ability to rely on an egocentric strategy to identify food locations. Trials were also divided into either under a local visual cue condition or a spatial relational condition involving distant environmental cues. Both hippocampal-lesioned and control monkeys discriminated the food locations in the presence of local cues. In the absence of local cues, control monkeys discriminated the food locations, whereas hippocampal-lesioned monkeys were unable to do so. The results demonstrate that the monkey hippocampal formation is critical for the establishment or use of allocentric spatial representations and that selective damage of the hippocampus prevents spatial relational learning in nonhuman primates.

There are also some unconventional mazes for nonhuman primates, such as a large virtual space in a virtual navigation task (Hori et al., 2005), a visual maze (Crowe, Chafee, Averbeck, & Georgopoulos, 2004) and a V-shaped maze (McDermott & Hauser, 2004). However, these mazes can at its best test spatial memory but not the type of cognitive function (e.g. learning of new spatial relations) lost in subjects with hippocampal damage. Furthermore, it came to our attention that some other more sophisticated monkeys maze models are being developed (Zhang et al., 2008), which will be valuable for future studies involving spatial learning and memory. All of these studies were designed more to assess spatial relational learning but less on exploratory behaviours per se, and mainly focused on selective hippocampal lesions. As I have contended somewhere else (Kwok, in press), fornix transection might produce different functional deficits from selective hippocampal lesions. In view of these considerations, we conducted an exploratory study with unrestrained monkeys, either with and without fornix lesions, in an enclosed chamber, in which they were allowed to explore the environment freely.

To facilitate this my colleague and I designed and built a novel octagonal test chamber within which unrestrained monkeys were able to freely ambulate around (Figure 2). At the start of this experiment the chamber was a completely novel environment to all of the experimental animals and for six consecutive days we observed and recorded the exploratory behaviour of a control group and a group of fornix transected macaques by means of an automated movement tracking system. We hypothesised that if the role of the hippocampus in exploration and spatial learning is conserved between rodents and primates, then we would observe that macaques with fornix transection would continue to explore the chamber at a similar rate across days, whereas the control macaques might show reduced exploratory behaviour as the chamber becomes an increasingly familiar environment. Furthermore, we also manipulated the degree of visual symmetry of the chamber between days to assess whether any such effect would only be seen on days where the directionality within the chamber was made explicit by an addition of asymmetrical visual cues.

Readers may be reminded that all the studies reviewed in a recent chapter (Kwok, in press, also by Nova Science Publishers) sought to assess one (or maybe two) very specific aspect of memory functions, and mostly involved small-scale environments in which monkeys responded by reaching, rather than by traveling to, different locations in space. Recognising that a holistic picture of normal functioning of an organism will be better shown with testing conducted under more naturalistic, or at least larger and immersive, environments, I detailed here the exploratory behaviours in respect to the effects of fornix transection on locomotion, exploration and learning in a group of macaque monkeys.

SUBJECTS

Six male cynomolgus monkeys (*Macaca fascicularis*) took part in this experiment. Their mean weight at the start of behavioural testing was 6.4 kg, and their mean age was 5 years and 3 months. All six monkeys had identical pre- and post-operative experience in concurrent discrimination learning tasks in series of experiments that were carried out before the present study began (M J Buckley, Wilson, Kwok, & Gaffan, 2005). They were housed together in a group enclosure, excepting two, who were housed together as a pair; all had automatically regulated lighting and with water available ad libitum.

SURGERY

Three of the six monkeys had received bilateral fornix transection (group FNX) and the remaining three were unoperated controls (group CON). All procedures were carried out in compliance with the United Kingdom Animals (Scientific Procedures) Act of 1986. The operations were performed in sterile conditions with the aid of an operating microscope, and the monkeys were anesthetized throughout surgery with barbiturate (5% thiopentone sodium solution) administered through an intravenous cannula. A D-shaped bone flap was raised over the midline and the left hemisphere. The dura mater was cut to expose the hemisphere up to the midline. Veins draining into the sagittal sinus were cauterized and cut. The left hemisphere was retracted from the falx with a brain spoon. A glass aspirator was used to make a sagittal incision no more than 5 mm in length in the corpus callosum at the level of the interventricular foramen. The fornix was sectioned transversely by electrocautery and aspiration with a 20 gauge metal aspirator insulated to the tip. The dura mater was drawn back but not sewn, the bone flap was replaced, and the wound was closed in layers. The operated monkeys rested for 11 – 14 days after surgery before beginning postoperative training. Unoperated control monkeys rested for the same period of time between preoperative and postoperative training.

HISTOLOGY

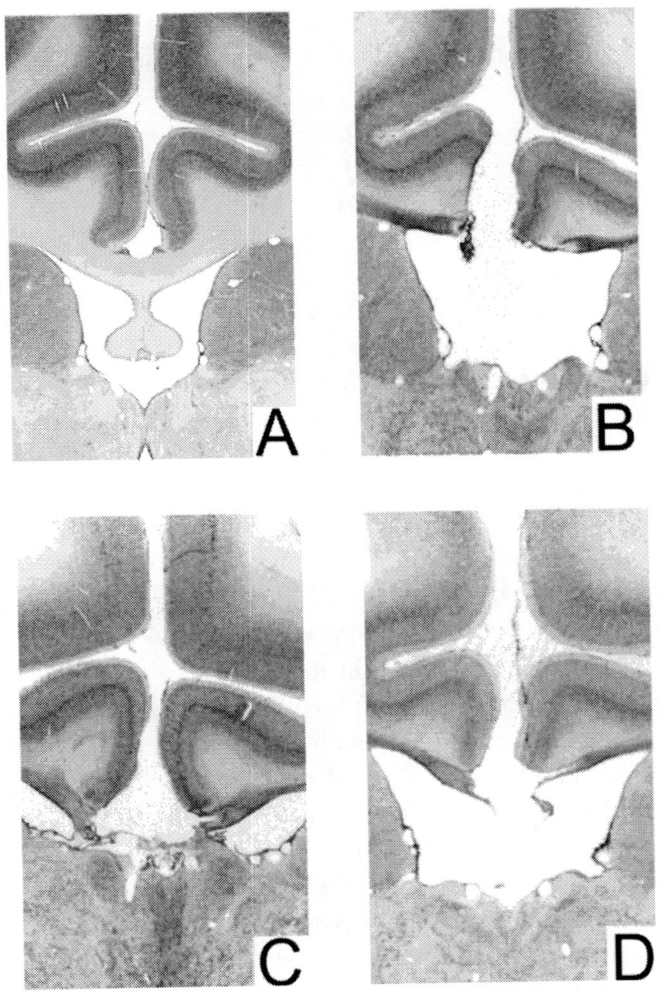

Figure 1. (A) Coronal section from the brain of a normal unoperated macaque just posterior to the level of the interventricular foramen; (B, C, D) coronal sections from the brains of three fornix transected monkeys showing that the fornix transection was complete (the anterior-posterior level of the fornix transection varies between these monkeys depending at which level the fornix was cut through a small hole made in the corpus callosum at that level).

At the conclusion of this and a number of further tasks that involved concurrent object discriminations (Wilson, Charles, Buckley, & Gaffan, 2007), visuospatial associative discriminations (Sze Chai Kwok & Mark J. Buckley, 2010), visuovisual associative learning (Kwok & Buckley, 2009), and long-term retention memory (Sze Chai Kwok & Mark J Buckley, 2010), the animals with fornix transection were deeply anaesthetised, then perfused through the heart with saline followed by formol-saline solution. The brains were blocked in the coronal stereotaxic plane posterior to the lunate sulcus, removed from the skull, and allowed to sink in sucrose-formalin solution. The brains were cut in 50 μm sections on a freezing microtome. Every fifth section was retained and stained with cresyl violet. Microscopic examination of the stained sections revealed in every case a complete section of the fornix (see Figure 1, panels B, C and D) with no damage outside the fornix except for the incision in the corpus callosum as described in the surgical procedures and at most, only slight damage to the most ventral part of the cingulate gyrus at the same anterior-posterior level in only one hemisphere of one animal (Figure 1, panel B). A coronal section of a normal control monkey's brain with an intact fornix is also shown for comparison (Figure 1, panel A).

Ambulatory Apparatus

The ambulatory apparatus was a symmetrical eight sided chamber with opposing sides of either 128 cm or 120 cm long (see Figure 2, panel 1 for a diagrammatic plan). The length across the chamber was 3.2 m and the total floor area was approximately 7.4 m^2. Four touchscreens (each with a visible screen area of 20 inches diagonally) were embedded in four of the walls for the purpose of displaying stimuli and registering the monkeys' responses to such stimuli (Figure 2, panel 1). An infra-red camera equipped with a wide angle lens was mounted centrally on the ceiling of the chamber (2.1 m above floor level) and was used with motion analysis software to continually record the movements of monkeys during the experiment. The walls and the floor of the chamber were coated with a durable waterproof material colored white and light yellow respectively upon which the dark coat of the monkey always stood out (due to differences in contrast) in the video images, thereby facilitating software analysis of the animal's motion in the captured video. In addition, four separate infra-red CCTV cameras equipped with infra-red light sources were also mounted high upon the ceiling in a symmetrical manner which allowed each one to capture a direct view of one of the four touchscreens, enabling the experimenter to observe the monkeys' screen touching behaviours from outside of the closed arena. Four extraction fans were installed, located on the top part of four of the wall panels to provide adequate ventilation within the enclosed arena during the testing session and the noise which these four fans generated while operated also acted as a mask of extra-chamber noise. One wall contained a large hinged door through which the experimenter could enter or leave the chamber between sessions to clean the chamber (three other walls contained a 'false' door to maintain visible symmetry). This large door (when closed) also contained within it a smaller door that could be raised or lowered to allow the monkey to enter or leave the chamber via a transport cage which could be securely abutted to the outside of the chamber at this point. When closed, the rectangular outline of this smaller door could be discerned from the inside and so the outlines of three 'false' small doors were drawn on three other walls to maintain visible symmetry. Altogether, a convincing visually-symmetrical environment was achieved within which human subjects quickly became disorientated after walking around for a few moments. However, on certain days, a non-symmetrical environment was created with the addition of visual aids to provide salient cues to directionality. A dramatic landscape poster (91 x 60 cm) was positioned above one touchscreen, a picture of a bird (30 x 27 cm) was positioned above another, and a 60W lamp shone through the extraction fan hole above a third screen (the effect of this lamp was to allow light to flood into the chamber from one point only in the room, akin to an artificial 'sun'). The asymmetrical environment was set up exactly the same way on each of these asymmetrical days. A mechanism to deliver food rewards to the monkey inside the chamber was incorporated by means of four white food pellet bowls positioned to the lower left hand side of each touchscreen; food rewards could be delivered from dispensers installed outside the chamber via a tube system connected to these bowls. However, these pellet dispensers (and the touchscreens themselves) were not used in this study.

PROCEDURE

For each session, a monkey was brought to the testing room in a wheeled transport cage, which abutted a sliding door of the apparatus, through which the monkey was introduced into the chamber. Each exploratory session lasted 20 minutes in which the animal was free to ambulate and explore the chamber at will. At the end of the session, the monkey left the chamber to be fed in its transport cage before being returned to its home cage. This procedure was repeated on 6 successive days, of which days 1, 3 and 5 were carried out in the symmetrical environment whereas days 2, 4 and 6 were carried out in the non-symmetrical environment. During the course of this study monkeys were given their daily food ration at the end of the daily session.

Movement Tracking and Arena Design

Each session was captured digitally by an overhead camera attached to a computer. TopScan software 1.0 (Clever Sys. Inc. Virginia, VA, 20190, USA) was used to determine the monkey's position and locomotion at a capturing speed of 30 frames/second. The video file was then compressed and analyzed within TopScan to generate the trace paths, cumulative traveled distance, number and duration of stops, traveling patterns and proportion of time spent in each of 12 designated zones. Figure 2 (panel 2) depicts each of the 12 zones used for analysis: the floor of the chamber was divided into 4 quarters (with a touchscreen centered in each) and each of these was subsequently split up into three smaller areas. This particular zoned design allowed us to assess which screen the monkey was closest to at any point in time and to observe whether or not a monkey had any predilection to spend time in particular areas within the chamber.

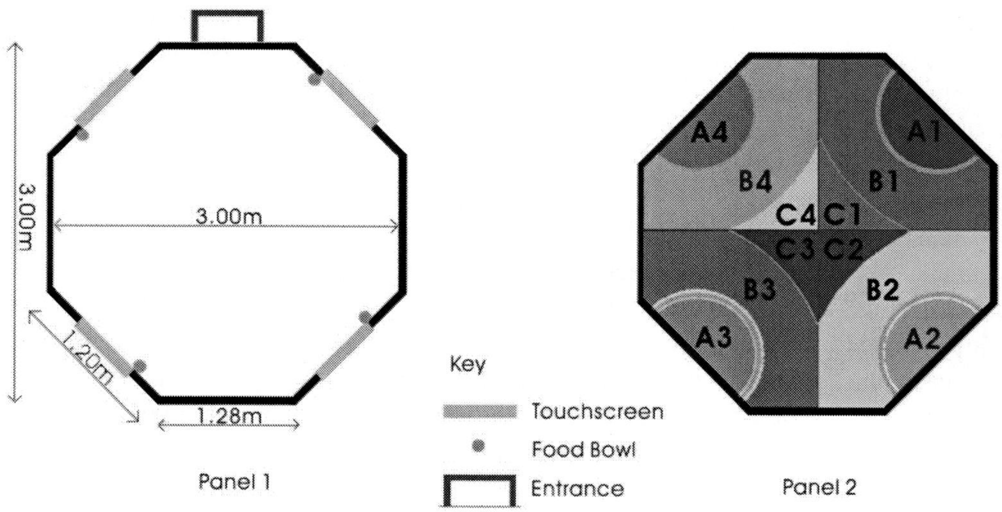

Figure 2. Panel 1. A plan of the ambulatory apparatus showing its dimensions and the positions of internal features, such as four touchscreens embedded in four of the walls, the four white food pellet bowls positioned next to the touchscreens, and the door through which a monkey enters into the chamber. Panel 2. A topographic representation of the apparatus arena (consisting of 12 areas).

Behavioural Measures

We recorded a number of behavioural measures in both the CON and FNX groups, across six consecutive days of testing. However on days 1, 3, and 5 we presented no visual cues to symmetry whereas on days 2, 4, and 6 we included the same set of cues to make the room visually asymmetrical. The following measures were used to examine the monkeys' exploratory movements on each daily session which lasted a total of 20 minutes:

[1] *Cumulative distance*. The exploratory paths taken by each monkey on each day were reconstructed and measured in length. The cumulative distances traveled by each monkey on the first day were also examined and analyzed within five consecutive four-minute epochs.

[2] *Number and duration of stops*. A stop was defined by a ratio of displacements of an animal's body between two consecutive time frames according to TopScan's motion measure function. An animal was classified as static (stopped) when the motion measure was below a value of 0.5. The duration of total stops was also measured.

[3] *Designated zones and screen quadrants*. The floor was divided into four main quadrants, each of them representing the area in front of a screen (i.e. screen quadrants 1, 2, 3 and 4) and each of these screen quadrants were further divided into three smaller zones (i.e. an 'outer' Zone A, an 'intermediate' zone B and an 'inner' zone C). Figure 2 (panel 2) shows the layout of each of the twelve zones. The proportion of time spent in each screen quadrant and zones were analyzed.

[4] *Trace paths and thoroughness of exploration*. A plan of chamber floor was divided into 244 individual squares by means of a regular grid superimposed upon the plan.

An average monkey body occupied roughly 10 squares of the grid. Trace paths of monkeys' movements were generated in accordance to the center of a monkey's body mass (computed by TopScan software) and each trace of exploration was then analysed with reference to this grid to determine the number of squares on the grid that were crossed either i) never, ii) once, or iii) more than once, this was done independently for every session of every monkey to examine the thoroughness of each daily exploration with respect to the proportion of floor-space visited.

RESULTS

Behavioural Results

Both the control and the fornix groups were active in that they locomoted extensively throughout the chamber during six 20-minute test sessions. Behaviour consisted of mainly locomotor progressions punctuated with periodic stops. Locomotion was mainly directed to the periphery of the chamber, but also included occasional trips across the center of the chamber. Comparisons between the control group and the fornix group across a wide range of behavioural measures showed that the two groups did not differ greatly in their behavioural profile across the six exploratory sessions.

Cumulative Distance

A) Across Six Daily Sessions

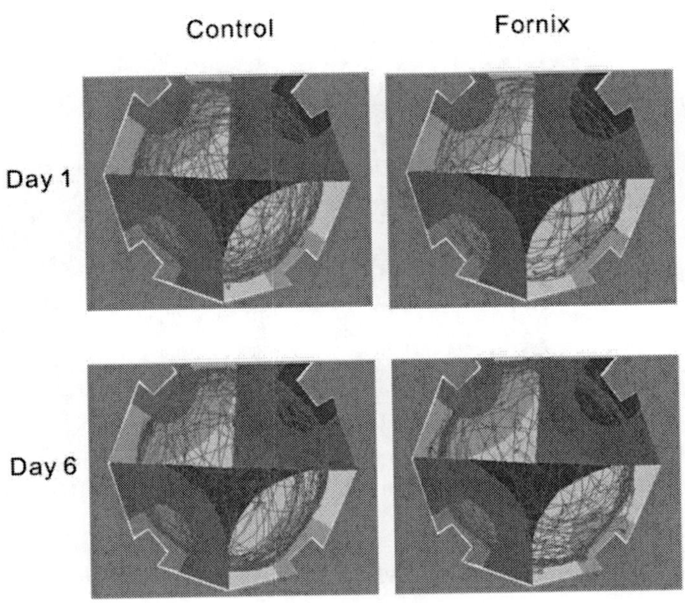

Figure 3. Traces of the exploratory paths followed by a representative control monkey (left panel) and a fornix lesioned monkey (right panel) on the first (day 1) and last day (day 6) of testing.

The mean distances traveled per 20-minute session averaged across the six sessions were 502 m for the CON group (range 359 m to 587 m) and 340 m for the FNX group (range 129 m to 621 m). An ANOVA of the average distance traversed per day showed that there were no significant differences between groups [F (1, 4) < 1]. A repeated measures ANOVA of the distance traveled on each of the six days of testing showed that there were also no significant changes in distance traveled by the groups across successive days [Session: F (5, 20) = 2.10, *p* > 0.1; Group*Session: F (5, 20) = 1.379, *p* > 0.1, and the linear trend component of this interaction was also insignificant] (see Figure 3 for examples of trace paths on the first and last days). An ANOVA considering the within-subjects differences between the mean daily distance traveled on symmetrical versus non-symmetrical days also showed no significant differences between groups [F (1, 4 < 1].

B) Across Epochs within the First Test Session (Day 1)

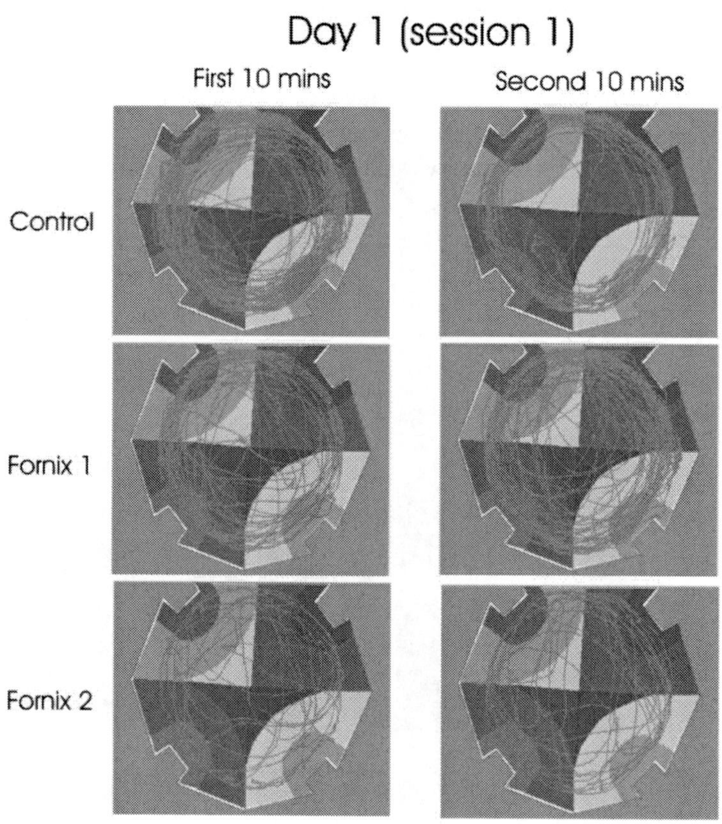

Figure 4. Traces of the exploratory paths followed by a control monkey (top panel) and two fornix lesioned monkeys (middle and bottom panels) in the first and second halves (10 minutes each) on the first day of testing.

As there was considerable individual variation in the mean distances traveled in the first epoch of the first testing session (Day 1) we conducted a repeated measures ANOVA on the logarithmic transformed distance traveled in five consecutive epochs within Day 1. This analysis showed that there was a significant difference between the groups in the within-subjects changes in distance traveled across these epochs [Group*Epoch: $F (4, 16) = 3.601$, $p = 0.028$] (see Figure 4 for example trace paths of animals from each group in the first and last half of session 1). Further, the linear trend component of this interaction was also significant [$F (1, 4) = 10.468$, $p = 0.032$] reflecting the fact that the difference between groups was apparent in the early epochs but not the latter epochs; indeed, despite marked individual differences in distance traveled in the first epoch, only the CON group animals showed a consistent tendency to travel less in later epochs (see Figure 5 for regression lines of the mean changes in distance traveled by each group across the five epochs of the first session).

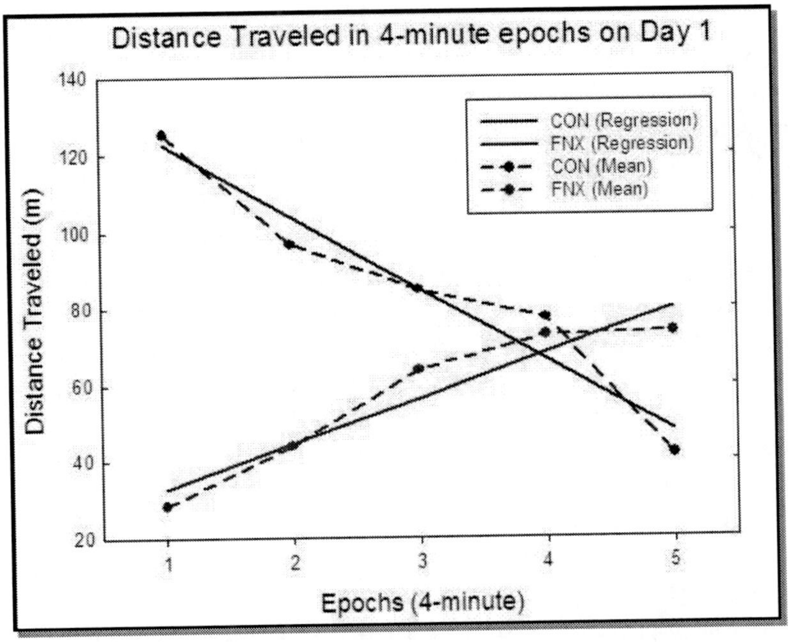

Figure 5. A graph showing the regression lines of the distances traveled in five 4-minute epochs by the CON and FNX groups on Day 1.

Stops and Duration of Stops

The mean numbers of stops averaged across the six sessions were 193 stops for the CON group (range 128 to 270) and 361 stops for the FNX group (range 234 to 536). An ANOVA of the average number of stop per day showed that there was no significant differences between groups [$F (1, 4) = 2.888$, $p > 0.1$]. A repeated measures ANOVA of the number of stops on each of the six days of testing showed that there were also no significant changes in the number of stops by the groups across successive days [Session: $F (5, 20) < 1$; Group*Session: $F (5, 20) < 1$, and the linear trend component of this interaction was also insignificant]. An ANOVA considering the within-subjects differences in the mean daily number of stops on symmetrical versus non-symmetrical days also showed no significant differences between groups [$F (1, 4) < 1$].

The mean duration of stops averaged across the six sessions were 271 seconds for the CON group (range 195 s to 391 s) and 429 seconds for the FNX group (range 184 s to 575 s). An ANOVA of the average duration of stops per day showed that there was no significant differences between groups [$F(1, 4) = 1.319$, $p > 0.1$]. A repeated measures ANOVA of the duration on each of the six days of testing showed that there were significant changes in the duration of stops across successive days [Session: $F(5, 20) = 9.709$, $p < 0.0001$] but there was no significant effect between groups [Group*Session: $F(5, 20) = 1.232$, $p > 0.1$, and the linear trend component of this interaction was also insignificant]. An ANOVA considering the within-subjects differences in the mean daily duration of stops depending upon whether the day was symmetrical or non-symmetrical also showed no significant differences between groups ($F(1, 4) = 3.039$, $p > 0.1$).

Zone Preference

The mean percentage of time spent inside the designated zones (see Figure 2, panel 2 for the division of zones) averaged across all of the six sessions were: zones A $^{1\text{-}4}$ (33.1% for CON and 33.0% for FNX); zones B $^{1\text{-}4}$ (63.6% for CON and 60.6% for FNX) and zones C $^{1\text{-}4}$ (3.3% for CON and 6.4% for FNX) respectively. ANOVAs showed that there were no significant differences between groups in the proportion of time which they spent in each of the different zones averaged across all days: zones A [$F(1, 4) < 1$], zones B [$F(1, 4) < 1$], and zones C [$F(1, 4) = 4.836$, $p = 0.09$] respectively. A repeated measures ANOVA indicated that there were some significant changes in the proportion of time spent by the animals in different zones across successive days of testing. The proportion of time spent by the monkeys in zones B varied significantly between days [Session: $F(5, 20) = 3.550$, $p = 0.018$] but there was no indication that the patterns of exploration exhibited by the two groups differed in this regard [Group*Session: $F(5, 20) < 1$]. As for other zones (zones A and zones C), there were no significant changes in zone preference across successive days and no significant changes between groups differences. ANOVAs considering the within-subjects mean differences in the percentage of time spent inside each of the zones depending upon whether the day was symmetrical or non-symmetrical also showed no significant differences between groups [Zone A: $F(1, 4) = 4.192$, $p > 0.1$; Zone B: $F(1, 4) < 1$; and Zone C: $F(1, 4) < 1$].

Screen Quadrant Preference and Effects of Symmetry

The mean percentage of time that the CON and FNX monkeys spent inside each of the screen quadrants (see Figure 2, panel 2) were as follows: for quadrant 1 (30.5% for CON and 24.7% for FNX); for quadrant 2 (19.9% for CON and 23.0% for FNX); for quadrant 3 (21.1% for CON and 25.8% for FNX); and for quadrant 4 (28.5% for CON and 26.4% for FNX) respectively. A repeated measures ANOVA of the mean percentage of time spent in each of the four quadrants (with 4 levels of the within-subjects factor quadrant, 2 levels of the within-subjects factor symmetry, and one between-subjects factor group) indicated that quadratic component of the group x quadrant interaction approached significance [$F(1, 4) = 5.646$ $p = 0.076$]. This, together with our observations that there were marked individual differences in preferences to quadrants and the feature of our design which ensured differential availability of different kinds of cues to each quadrant in asymmetrical days prompted us to examine the data in closer detail. Thus, even despite individual differences in preferences we discerned that the CON group appeared on average to prefer to spend proportionally more time in the two quadrants on either side of, and therefore equally closest to, the exit (Figure 2, panel 2). A t-test confirmed that averaged over the 6 days the CON group's summed preference scores for these two 'exit' quadrants as opposed to the other two 'non-exit' quadrants was indeed significantly greater than that of the FNX group ($t = 2.379$, $df = 4$, $p = 0.038$, 1-tailed). Furthermore, only one of the exit quadrants (quadrant 1) was illuminated by the artificial sun while the other (quadrant 4) remained in relative darkness. Although individual subjects exhibited varying preferences to quadrants with a single 'attractive' feature (exit proximity or lightness) examination of the data showed that a clear consensus in preference emerge in the CON group towards spending proportionally more time in the zone with both alluring features present (Figure 6). Indeed, an ANOVA considering the within-subjects differences in the percentage of time spent inside each of the four quadrants depending upon whether the day was symmetrical or non-symmetrical showed that the CON group tended to increase the amount of time spent within a particular quadrant (quadrant 1) when the chamber was made asymmetrical but that the FNX group showed no such bias on asymmetrical days [$F(1, 4) = 14.266$, $p = 0.019$]. No such effect was found with the other three quadrants (all F's $(1, 4) < 1$).

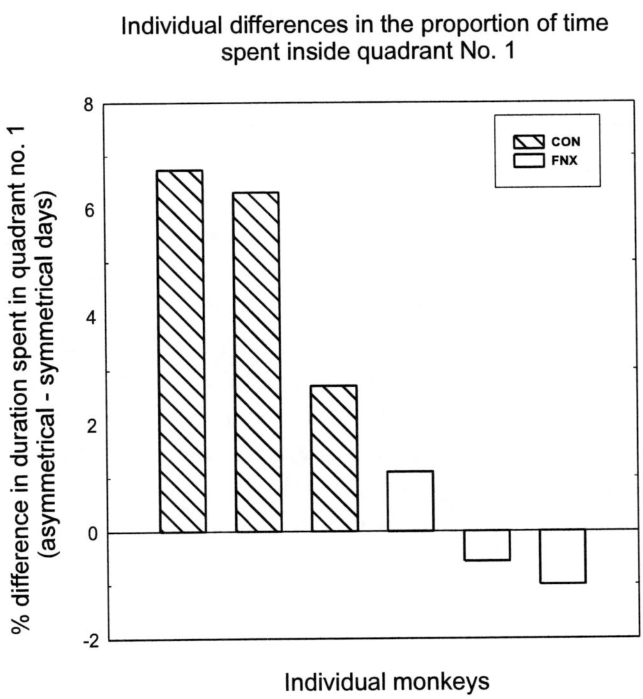

Figure 6. A graph showing the differences in the mean percentage of time (asymmetrical days – symmetrical days) spent within quadrant no. 1 for six individual monkeys.

Proportion of Arena Visited

The average proportions of the arena visited by the monkeys averaged across the six sessions were 92.5% for the CON group (range 84.6% to 98.4%) and 94.2% for the FNX group (range 87.9% to 98.7%). An ANOVA of the percentage of squares visited showed that there were no significant differences between groups in all three measures: i) squares never visited, $F(1, 4) = 0.106$, $p > 0.1$; ii) squares visited only once, $F(1, 4) = 0.053$, $p > 0.1$; and iii) squares visited more than once, $F(1, 4) < 1$. A repeated measures ANOVA of the proportion of the arena visited on each of the six days of testing showed that there were no significant changes in the proportion of the arena visited by the groups across successive days [Session: $F(5, 20) = 0.839$, $p > 0.1$; Group*Session: $F(5, 20) < 1$, and the linear trend component of this interaction was also insignificant]. ANOVAs of the within-subjects differences in the mean proportion of the arena visited on symmetrical versus non-symmetrical days also showed no significant differences between groups in all three measures [squares never visited: $F(1, 4) < 1$; squares visited only once: $F(1, 4) = 1.561$, $p > 0.1$; and squares visited more than once: $F(1, 4) < 1$].

Individual Differences

Despite the overall pattern of the main effects of the behavioural measures, there were some striking individual differences in the behaviour pattern of FNX monkeys. Similar to hippocampal rats [(O'Keefe & Nadel, 1978), Table A14], the effect of fornix lesion on hyperresponsivity in our FNX monkeys was also not invariant. For instance, the most and the least active monkeys of all were both from the FNX group. In the measure of distance traveled, the least active monkey (FNX3) traveled merely 129 meters on average across six sessions, whereas the most active one (FNX1) traveled 621 meters (compared to 421 meters overall mean of all six monkeys). Evidence of this bipolar activity level within FNX group also manifested itself in other measures such as the number of stops and the total duration of static state. In the aspect of traversing patterns, one lesioned monkey (FNX2) traversed distinctly from others (e.g. FNX1) in a manner that it did not explore in a circling pattern around the periphery of the chamber but making many crosses and shortcuts across the chamber instead (Figure 4, contrasting middle and bottom panels). In order to minimise the variances due to individual differences, many of our analyses above have taken individual differences into account by considering within-subjects measures such as change in individual activity levels over epochs, and over days, and by using comparisons of within-subjects activities on symmetrical versus asymmetrical days.

Conclusion

Following Clark et al. (2005)'s findings of intact exploratory behaviour in hippocampal lesioned rats on an open circular table, the present study tested the prediction that transection of the fornix in monkeys would not abolish their exploratory behavioural patterns in a novel environment. On 6 successive days, control and fornix transected monkeys were introduced into an octagonal chamber in which their locomotion and movements of exploration were assessed according to a number of measures. Both CON and FNX monkeys visited, punctuated by stops, the majority of the floor-space of the chamber in each of the 20-minute sessions. CON and FNX monkeys were surprisingly similar on most measures of exploratory movements: including the cumulative distance traveled, the proportion of the floor-space of the chamber traversed, the number of stops and the amount of time spent still. We observed that FNX monkeys were as active as controls, but showed no signs of hyperactivity relative to controls, in locomoting within the chamber across all 6 days. The similarity of activity level between groups is taken as evidence that FNX monkeys are not un-motivated relative to the CON monkeys. The present results confirm that locomotor ability and motivation for exploring in a novel environment are well preserved in macaque monkeys with fornix lesions. This strengthens arguments (Gaffan, 1998) against the claim made by Whishaw and Jarrard (1996) that the hippocampal system, rather than contributing to learning per se, might instead mediate integrative processes providing movements that lead to learning. Our findings indicate that disrupting the normal function of the hippocampal system in non-human primates, by means of fornix transection, has no effect upon these proposed motor processes. Thus, in so far as our behavioural measures assess the extent of exploratory behaviour, our results do not support the hypothesis put forward by O'Keefe and Nadel (1978) that, if subject to hippocampal system disruption, animals would cease to display exploratory behaviours.

Clark et al. (2005) showed that whilst unoperated control rats exhibited 'habituation' (in terms of sustaining reactivity to salient aspects of a testing situation) over successive days of testing in a novel environment, rats with hippocampal lesions in contrast did not. In our experiment in which macaques were similarly introduced into a novel environment, neither the CON nor FNX group showed signs of habituation over successive days as both groups manifested similar exploratory movements across the six daily test sessions. However, we did find evidence for habituation with regard to exploratory behaviour *within* the first test session as our CON group exhibited a consistent decrement in traveled distance within consecutive epochs of the first test session whereas our FNX group did not. The failure of our FNX group to display 'habituation' within the first session is consistent with previous evidence in rat studies that animals with hippocampal system disruption behave differently, with regard to sustaining reactivity to salient aspects of a testing situation (Clark, Hines, Hamilton, & Whishaw, 2005).

With regards to Clark et al. (2005)'s study, one could estimate a ratio of the body mass of rats (250 – 300 g on average) to the area of the circular table (5.1 m^2) that could be explored by the rats in that study. In order to allow our macaques to explore an environment with the same body mass to area ratio we would have had to build an arena measuring approximately 16 times larger than the 7.4 m^2 of floor space available in our current arena. Not only was our novel environment relatively smaller in scale than the novel environment given to rats by Clark and colleagues (2005), it may also be considered to be relatively simplistic in terms of the extent of distinguishing visual features present a real naturalistic environment for macaques, say in a forest. Therefore, it is at least possible to speculate that the relatively small size and low complexity of our environment may have been a contributing factor towards our CON macaques' abilities to habituate within a very short period of time (within the first 20 minutes) so that no further decrements in measurements could be observed across subsequent days. However, it is also interesting to note that preliminary studies have indicated that when rats are allowed to explore 'infinite' virtual environments will freely locomote (or 'explore') vast environmental ranges covering up to 220 meters in a single 10-minute session (Hölscher, Schnee, Dahmen, Setia, & Mallot, 2005) suggesting that the environmental ranges given to rats in experimental contexts are similarly restrictive in allowing the expression of natural exploratory tendencies. Thus, while our data is at least suggestive that non-human primates may be able to habituate to visual environmental cues at a faster rate than rodents, in the absence of any means to evaluate the relative efficacy of perceptual cues to directionality employed by these two different studies, with different species, and with different environmental ranges, this interpretation must remain speculative at present. Furthermore, other potentially important aspects of species differences, such as the relative importance of visual exploration in monkeys compared to olfactory and tactile exploration in rats, would also need to be taken into account before making any strong claims about general species differences in habituation rate.

We found one further measure that distinguished the behaviour of the unoperated control macaques from that of those with fornix transection in our novel experimental arena. Unlike fornix transected animals, control animals showed stronger preferences for spending time in certain regions of the environment. In particular we found that control animals exhibited a mild but statistically significant preference for spending relatively more time in one particular quadrant of the arena (quadrant number 1) on those days in which asymmetrical visual cues gave a stronger sense of directionality to the environment whereas fornix transected animals did not show such a preference on either symmetrical or asymmetrical days. I speculate that the preference for quadrant number 1 might have resulted from a combination of factors. Firstly, with the aid of visual landmarks on asymmetrical days, the CON group may have had a greater capacity to remember from which direction they were introduced into the arena and therefore might have been better able to keep track of the location of the entrance/exit to the arena. Quadrants 1 and 4 happened to be the two quadrants located right next to the exit and all things being equal monkeys might have preferred to locate themselves in and around these quadrants with a view to expediting their exiting from the arena when the opportunity arose (overall, our analyses confirmed that control monkeys indeed expressed a preference for locating themselves within these two quadrants, but that this was not the case in fornix transected animals). However, the chamber was also set up with an 'artificial sun' shining light into the arena from above quadrant 4 and which primarily illuminating the other three quadrants. All things being equal, monkeys may have preferred to spend more time in quadrants 1 – 3 either because these positions afforded a view of the source of the light form outside the arena, or perhaps because the relative brightness of these zones afforded a greater sense of security while in an unfamiliar environment. The combination of these two factors may explain why quadrant 1 was much preferred over other quadrants for normal monkeys on asymmetrical days as it was the only quadrant which presented both potential benefits.

There are a number of potential reasons for why fornix lesioned macaques may have failed to habituate and to develop regional preferences in response to salient visual cues to directionality available on asymmetrical days. One potential explanation is that our lesioned animals might simply have been impaired at recognising non-spatial or spatial visual features. However we judge this as unlikely as monkeys with fornix transection have been shown to be completely unimpaired at discriminating objects in the context of trial-unique object-recognition memory tasks (Charles, Gaffan, & Buckley, 2004), and fornix lesioned macaques have also been shown to remain unimpaired at discriminating large numbers of spatial problems exposed to preoperatively (Mark J. Buckley, Charles, Browning, & Gaffan, 2004; M J Buckley, Wilson, Kwok, & Gaffan, 2005). Indeed, object and spatial recognition memory has been shown, under some circumstances, to even survive neurotoxic lesions directed towards the hippocampus and amygdala in macaques (Elisabeth A Murray & Mishkin, 1998). Moreover, differences in forgetting rate between CON and FNX groups are also deemed to be unlikely to account for the group differences because fornix transection in macaques was shown not to affect the rate at which over 100 visuospatial discrimination problems were forgotten even after an interval as long as 15 months (Sze Chai Kwok & Mark J Buckley, 2010).

On the other hand, the lack of selective habituation in the FNX group may reflect the necessity of an intact hippocampal system to build and update the internal representation of the spatial layout of an environment during exploration (Clark, Hines, Hamilton, & Whishaw, 2005; O'Keefe & Nadel, 1978). Indeed, some spatial learning theories stipulate that the hippocampal system contributes to exploratory behaviour indirectly by facilitating animals' acquisition of spatial information regarding their environment (Hines & Whishaw, 2005). While fornix transection has previously been shown to impair macaques' abilities to learn about scenes (Gaffan, 1994b), such scene learning tasks do not explicitly demand spatial learning. Nevertheless, other tasks that do explicitly tax spatial learning have been shown to be impaired after fornix transection (M J Buckley, 2005; Mark J. Buckley, Charles, Browning, & Gaffan, 2004; T J Bussey, Saksida, & Murray, 2005; Gaffan, 1994a; Gaffan & Harrison, 1989; E. A. Murray, Davidson, Gaffan, Olton, & Suomi, 1989). Furthermore, as recent experiments by Charles et al. (2004) and Brasted et al. (2003) have indicated that an intact fornix is also required to support various kinds of temporal information processing tasks, thus one can infer that fornix lesions impairs learning about spatial-temporal context. I propose that a general impairment in learning about context could have impeded the ability of FNX animals to develop location preferences as well as their ability to habituate to the novel environment at the same rate as controls.

Nonetheless, we still cannot completely rule out that the behavioural differences between our FNX and CON monkeys might be attributable to underlying deficits of a perceptual nature. According to the MTL (Medial Temporal Lobe) Memory System theory, the MTL contributes exclusively to mnemonic processing (Squire & Zola-Morgan, 1991). However, more recently some authors have challenged this theory on account of evidence that some MTL structures, particularly the perirhinal cortex, contribute to perception [for recent reviews see: (M J Buckley, 2005; Mark J Buckley & Gaffan, 2006; Timothy J Bussey & Saksida, 2005; T J Bussey, Saksida, & Murray, 2005; Lee, Barense, & Graham, 2005; Elisabeth A Murray & Bussey, 1999)]. By this account, MTL structures are critical to higher order perceptual as well as mnemonic processes, and therefore the hippocampus might be involved in perception, of scenes and spatial information in particular. This would be consistent with recent new evidence demonstrating deficits in a variety of spatial processing tasks which place little or no demands on memory following hippocampal damage (Hartley et al., 2005; Lee, Barense, & Graham, 2005; Lee et al., 2005). Although in the context of the current study, we are unable to conclude whether the failure of the FNX group to habituate and develop location preference was caused either by mnemonic deficits, perceptual deficits or both, I advocate the view that a role for the hippocampus in spatial perception is not necessarily contradictory to a relational memory view of the hippocampus (H Eichenbaum, Otto, & Cohen, 1994). The hippocampal system could be involved in the binding of perceptually distinct items both for explicit long-term memory as well as for more immediate processing of the relationships among perceptually distinct elements of scenes or events that give rise to perceived spatio-temporal context.

In this study, although monkeys with fornix transection habituated to their new environment in a chamber at a slower rate than their normal counterparts during the first 20-

min session, their habituation caught up after the first testing session and matched the controls' level eventually. Since exploratory learning of an environment involves learning of a wide range of information, we cannot ascertain in that experiment whether the slower learning of the environment was centred on deficits in learning the visuospatial or just the spatial elements of the context, or in perceptual processing of the context, or other non-mnemonic factors such as differential responses to 'stress-inducing' features of the testing situation (Gray, 1982). The finding however suggests a possibility that the learning deficits caused by fornix transection are not always of an all-or-none manner. It might be a slowing in the initial phases wherein associative information can be learnt rapidly with an intact hippocampal system (Brasted, Bussey, Murray, & Wise, 2003; McClelland, McNaughton, & O'Reilly, 1995) as it has been theorised, from the perspective of connectionist modelling, that the hippocampus might underlie the most rapid learning phases of new associative information (McClelland, McNaughton, & O'Reilly, 1995). We have subsequently attempted to test this 'fast learning' idea more rigorously by training animals with or without fornix transection to acquire conditional visuospatial problems in another study (Sze Chai Kwok & Mark J. Buckley, 2010), and found that monkeys with fornix transection are impeded in acquiring this kind of problem. Added on to the fast learning impairments already known in visuomotor conditional problems (Brasted, Bussey, Murray, & Wise, 2003), that study provides new evidence to extend the scope of fast learning impairments to a new kind of problem, namely conditional visuospatial. Data from neurophysiological studies showing that neuronal activity changes relatively early in the hippocampus during the learning of location-scene associations (Wirth et al., 2003) also indicate a wider hippocampal role in subserving the process of fast learning of spatial associations.

In conclusion, in the present study FNX monkeys displayed a behavioural pattern that was very similar to that of control animals overall excepting for the fact that they failed to habituate to a novel environment in the same time frame as controls and that they failed to develop a predilection of a particular quadrant. We are able to conclude that motivation and locomotor ability remain intact after FNX transection and to reject the hypothesis that fornix lesions abolish exploratory behaviour per se. Based on the fact that FNX monkeys failed to habituate and to develop location preferences, I suggest that an intact fornix is necessary for habituation to a novel environment, as well as to build and update the internal representation of the spatial layout of an environment during exploration. Whether the underlying nature of this deficit is perceptual or mnemonic remains to be elucidated, and this is therefore a topic of considerable interest that future work should aim to address.

Acknowledgments

I thank Dr Mark J. Buckley; and with much love to Birgitta Oi Yan Tam.

References

Brasted, P. J., Bussey, T. J., Murray, E. A., & Wise, S. P. (2003). Role of the hippocampal system in associative learning beyond the spatial domain. *Brain, 126*, 1202-1223.

Buckley, M. J. (2005). The role of the perirhinal cortex and hippocampus in learning, memory, and perception. *Quarterly Journal of Experimental Psychology Section B, 58*(3/4), 246-268.

Buckley, M. J., Charles, D. P., Browning, P. G. F., & Gaffan, D. (2004). Learning and retrieval of concurrently presented spatial discrimination tasks: role of the fornix. *Behavioral Neuroscience, 118*, 138-149.

Buckley, M. J., & Gaffan, D. (2006). Perirhinal cortical contributions to object perception. *Trends in Cognitive Sciences, 10*(3), 100-107.

Buckley, M. J., Wilson, C. R. E., Kwok, S. C., & Gaffan, D. (2005, 13 Nov 2005). *Fornix transection results in a selective impairment of new learning of concurrent visuospatial discrimination problems; it has no effect upon retention, or the rate of forgetting, of large numbers of concurrently learned problems.* Paper presented at the 35th Society for Neuroscience Conference, Washington, DC.

Bussey, T. J., & Saksida, L. M. (2005). Object memory and perception in the medial temporal lobe: an alternative approach. *Current opinion of neurobiology, 15*(6), 730-737.

Bussey, T. J., Saksida, L. M., & Murray, E. A. (2005). The perceptual-mnemonic/feature conjunction model of perirhinal cortex function. *The Quarterly Journal of Experimental Psychology B, 58*(3-4), 269-282.

Charles, D. P., Gaffan, D., & Buckley, M. J. (2004). Impaired recency judgments and intact novelty judgments after fornix transection in monkeys. *Journal of Neuroscience, 24*(8), 2037-2044.

Clark, B. J., Hines, D. J., Hamilton, D. A., & Whishaw, I. Q. (2005). Movements of exploration intact in rats with hippocampal lesions. *Behavioural Brain Research 163*(1), 91-99.

Crowe, D. A., Chafee, M. V., Averbeck, B. B., & Georgopoulos, A. P. (2004). Neural activity in primate parietal Area 7a related to spatial analysis of visual mazes. *Cerebral Cortex, 14*(1), 23-34.

Easton, A., Parker, A., Derrington, A. M., & Parker, A. (2003). Behaviour of marmoset monkeys in a T-maze: comparison with rats and macaque monkeys on a spatial delayed non-match to sample task. *Experimental Brain Research, 150*, 114-116.

Eichenbaum, H. (2000). A cortical-hippocampal system for declarative memory. *Nature Review Neuroscience, 1*(1), 41-50.

Eichenbaum, H., Otto, T., & Cohen, N. J. (1994). Two functional components of the hippocampal memory system. *Behavioral and Brain Sciences, 17*, 449-518.

Gaffan, D. (1994a). Dissociated effects of perirhinal cortex ablation, fornix transection and amygdalectomy: evidence for multiple memory systems in the primate temporal lobe. *Experimental Brain Research, 99*, 411-422.

Gaffan, D. (1994b). Scene-specific memory for objects: A model of episodic memory impairment in monkeys with fornix transection. *Journal of Cognitive Neuroscience, 6*(4), 305-320.

Gaffan, D. (1998). Idiothetic input into object-place configuration as the contribution to memory of the monkey and human hippocampus: a review. *Experimental Brain Research, 123*, 201-209.

Gaffan, D., & Harrison, S. (1989). Place memory and scene memory: effects of fornix transection in the monkey. *Experimental Brain Research, 74*, 202-212.

Gray, J. A. (1982). *The neuropsychology of anxiety*. Oxford, UK: Clarendon Press.

Hampton, R. R., Hampstead, B. M., & Murray, E. A. (2004). Selective hippocampal damage in Rhesus monkeys impairs spatial memory in an open-field test. *Hippocampus, 14*, 808-818.

Harley, C. W., & Martin, G. M. (1999). Open field motor patterns and object marking, but not object sniffing, are altered by ibotenate lesions of the hipppocampus. *Neurobiology of Learning and Memory 72*, 202-214.

Hartley, T., Bird, C. M., Chan, D., Cipolotti, L., Husain, M., Vargha-Khadem, F., et al. (2005). The four mountains test: hippocampal damage selectively impairs topographical processing of spatial scenes over very short delays. *Society for Neuroscience Abstracts*, 191.114.

Hines, D. J., & Whishaw, I. Q. (2005). Home bases formed to visual cues but not to self-movement (dead reckoning) cues in exploring hippocampectomized rats. *European Journal of Neuroscience, 22*(9), 2363-2375.

Hölscher, C., Schnee, A., Dahmen, H., Setia, L., & Mallot, H. A. (2005). Rats are able to navigate in virtual environments. *Journal of Experimental Biology, 208*, 561-569.

Hori, E., Nishio, Y., Kazui, K., Umeno, K., Tabuchi, E., Sasaki, K., et al. (2005). Place-related neural responses in the monkey hippocampal formation in a virtual space. *Hippocampus, 15*(8), 991-996.

Kwok, S. C. (in press). Mnemonic role of the fornix: insights from the macaque monkey brain. In *Amnesia: Causes, Diagnosis and Treatments*. New York: Nova Science Publishers.

Kwok, S. C., & Buckley, M. J. (2009). Fornix transected macaques make fewer perseverative errors than controls during the early stages of learning conditional visuovisual discriminations. *Behavioural Brain Research, 205*(1), 207-213.

Kwok, S. C., & Buckley, M. J. (2010). Fornix transection selectively impairs fast learning of conditional visuospatial discriminations. *Hippocampus, 20*(3), 413-422.

Kwok, S. C., & Buckley, M. J. (2010). Long-term visuospatial retention unaffected by fornix transection. *Hippocampus, 20*(8), 889-893.

Lavenex, P. B., Amaral, D. G., & Lavenex, P. (2006). Hippocampal lesion prevents spatial relational learning in adult macaque monkeys. *Journal of Neuroscience, 26*(17), 4546-4558.

Lee, A. C. H., Barense, M. D., & Graham, K. S. (2005). The contribution of the human medial temporal lobe to perception: Bridging the gap between animal and human studies. *The Quarterly Journal of Experimental Psychology B, 58*(3/4), 300-325.

Lee, A. C. H., Bussey, T. J., Murray, E. A., Saksida, L. M., Epstein, R. A., & Kapur, N. (2005). Perceptual deficits in amnesia: Challenging the medial temporal lobe 'mnemonic' view. *Neuropsychologia, 43*, 1-11.

McClelland, J. L., McNaughton, B. L., & O'Reilly, R. C. (1995). Why there are complementary learning systems in the hippocampus and neocortex: Insights from the successes and failures of connectionist models of learning and memory. *Psychological Review, 102*(3), 419-457.

McDermott, J., & Hauser, M. (2004). Are consonant intervals music to their ears? Spontaneous acoustic preferences in a nonhuman primate. *Cognition, 94*(2), B11-B21.

Murray, E. A., & Bussey, T. J. (1999). Perceptual - mnemonic functions of the perirhinal cortex. *Trends in Cognitive Sciences, 3*(4), 142-151.

Murray, E. A., Davidson, M., Gaffan, D., Olton, D. S., & Suomi, S. (1989). Effects of fornix transection and cingulate cortical ablation on spatial memory in rhesus monkeys. *Experimental Brain Research, 74*, 173-186.

Murray, E. A., & Mishkin, M. (1998). Object recognition and location memory in monkeys with excitotoxic lesions of the amygdala and hippocampus. *Journal of Neuroscience, 18*(16), 6568-6582.

Nadel, L., & Hardt, O. (2004). The spatial brain. *Neuropsychology, 18*(3), 473-476.

O'Keefe, J., & Nadel, L. (1978). *The hippocampus as a cognitive map.* Oxford, UK Clarendon.

Redish, A. D. (1999). *Beyond the cognitive map: from place cells to episodic memory.* Cambridge Mass: MIT Press.

Squire, L. R., & Zola-Morgan, S. (1991). The medial temporal lobe memory system. *Science, 253*(5026), 1380-1386.

Whishaw, I. Q., Cassal, J. C., & Majchrzak, M. (1994). "Short-stops" in rats with fimbria-fornix lesions: evidence for change in the mobility gradient. *Hippocampus, 4*, 577-582.

Whishaw, I. Q., & Jarrard, L. E. (1996). Evidence for extrahippocampal involvement in place learning and hippocampal involvement in path integration. *Hippocampus, 6*, 513-524.

Wilson, C. R. E., Charles, D. P., Buckley, M. J., & Gaffan, D. (2007). Fornix transection impairs learning of randomly changing object discrimination. *Journal of Neuroscience, 27*(47), 12868-12873.

Wirth, S., Yanike, M., Frank, L. M., Smith, A. C., Brown, E. N., & Suzuki, W. A. (2003). Single Neurons in the Monkey Hippocampus and Learning of New Associations. *Science, 300*(5625), 1578-1581.

Zhang, B., Tan, H., Sun, N.-L., Wang, J.-H., Meng, Z.-Q., Li, C.-Y., et al. (2008). Maze model to study spatial learning and memory in freely moving monkeys. *Journal of Neuroscience Methods, 170*(1), 111-116.

In: Monkeys: Biology, Behavior and Disorders
Editor: Rachel M. Williams, pp 119-135
ISBN: 978-1-61209-911-8

Chapter 5

NEUROBIOLOGICAL CORRELATES OF BEHAVIORAL AND COGNITIVE PERFORMANCE IN NONHUMAN PRIMATES

***Gwendolen E. Haley*[1,2] *and Jacob Raber*[1,2,3]**
[1]Behavioral Neuroscience Department, Oregon Health and Science University
Portland, OR 97239
[2]Division of Neuroscience, Oregon National Primate Research Center
Beaverton, OR 97006
[3]Department of Neurology, Oregon Health and Science University
Portland, OR 97006

ABSTRACT

Animals in experiments are traditionally grouped by experimental treatment. Although this is a valuable way to differentiate the groups, alternatively, groups can be distinguished based on cognitive performance. Performance based analysis can yield valuable insights, corresponding to behavior and/or cognition, that might not otherwise be observed. As an example of such an analysis, we discuss a cohort of elderly female rhesus macaques who participated in a spatial food port maze navigational test. Circadian activity and pharmacological MRI (phMRI) were assessed in these monkeys *in vivo* and radioligand binding was assessed in post-mortem tissue. Based on cognitive performance in the spatial maze, the cohort of monkeys was divided into Good Spatial Performers (GSP) and Poor Spatial Performers (PSP). GSP animals were more active during the day and less active at night compared to PSP animals. In addition, GSP animals had a higher percentage change in blood-oxygen-level-dependent (BOLD) signal after a scopolamine challenge, a non-specific muscarinic receptor antagonist, compared to PSP animals. Post-mortem radioligand experiments demonstrated that hippocampal muscarinic type 1 (M_1) maximum receptor binding and receptor binding affinity, hypothesized to have an integral role in spatial learning and memory, was significantly greater in the hippocampus of GSP than PSP animals. In contrast to the hippocampus, M_1 receptor binding was not significantly different in the prefrontal cortex (PFC). Muscarinic type 2 (M_2) maximum receptor binding and receptor binding affinity were not significantly different between the two groups in either brain region. Finally, there were positive correlations between circadian measures and the percentage change in BOLD signal following the scopolamine challenge, as well as M_1 receptor binding measures. Thus, GSP animals sleep more and have anenhanced M_1 receptor function. These data demonstrate the close relationship between BOLD signal changes, circadian activity, and M_1 receptor binding parameters. Distinguishing groups based on cognitive or behavioral performance is valuable for studying neurobiological correlates of performance in nonhuman primates.

Keywords: behavioral groups, muscarinic receptors, phMRI.

INTRODUCTION

Traditionally, when designing experiments, animals are grouped based on an experimental manipulation. During the course of the experimentation, especially when assessing animal behavior or cognition, a variety of data is collected. An alternative way to examine group differences is a post-hoc type of analysis in which animals are grouped based on their behavioral or cognitive performance. For instance, post-mortem tissue analyses could conceivably present different results that could be related to a specific type of behavior or cognitive testing rather than the experimental condition. This approach can especially be successful when using a multidisciplinary approach to examine the animals. In this chapter, we will present and discuss such analysis using data from a previous study in which muscarinic receptor function was assessed *in vivo* and *in vitro* in post-mortem tissue from the same animals.

Traditional experiments separate groups based on an independent variable. For instance, when assessing age-related changes in a particular physiological area, distinct groups are traditionally based on age [1]. Likewise, groups can be based on genetic differences [2]. Alternatively, groups can be based on a specific treatment versus control [3, 4]. However, with behavioral or cognitive experiments, performance-based groups might demonstrate neurobiological differences with strong implications in translational research projects. As invasive diagnostic methods are used sparingly in the clinic, an idea of what physiological parameters are reflected in behaviors would be most valuable. Specifically, with regard to aging and age-related pathologies, understanding behavioral and cognitive representation could improve health care for the increased aging population.

Muscarinic acetylcholine receptors are associated with post-synaptic neurotransmission and memory function [5]. They are present at high concentrations in the prefrontal cortex (PFC) and the hippocampus [6] and have an integral role in spatial learning and memory in rodents, nonhuman primates, and humans [7-10]. Scopolamine, a nonspecific muscarinic receptor antagonist [11], impairs cognitive task performance in rats [12], dogs [13], rhesus monkeys [14, 15], and humans [16]. The most highly expressed muscarinic receptor in the PFC and hippocampus, regions critical for cognitive function, is the type 1 (M_1) subtype [7-10, 17-19]. The muscarinic type 2 (M_2) is the second most abundant muscarinic receptor in the PFC and hippocampus [20, 21] and might also contribute to cognitive function [22]. M_1 receptor deficient mice show impaired working memory and memory consolidation and M_2 receptor deficient mice show impaired passive avoidance task performance [23]. Nonhuman primate studies show a decrease in M_1 and M_2 receptor number with age [24]. Other reports indicate an age-related decrease in M_1 receptor affinity [25]. In patients with Alzheimer's disease, M_1 receptor agonists are proposed to increase cognitive function, increase neurogenesis, and decrease tau phosphorylation [26-29]. Together, these findings suggest that the M_1, and to a lesser extent M_2, receptors may play a key role in the regulation of cognitive function in the elderly.

Cognitive function is negatively impacted by disruptions in circadian activity, especially in the elderly [30, 31]. For example, circadian activity has been correlated with cognitive performance on a spatial task in old female rhesus monkeys [32]. The M_1 receptor might be involved in regulation of circadian activity [33, 34] and this in turn might contribute to the M_1 receptor-mediated cognitive effects. Neurotransmitter function can be readily assessed *in vivo* using pharmacological magnetic resonance imaging (phMRI). Although similar to functional magnetic resonance imaging (fMRI), which measures blood oxygen level dependent (BOLD) signal changes associated with cognitive performance, phMRI measures BOLD signal changes following exposure to pharmacological stimuli [35]. This technique has been successfully used in humans and nonhuman primates to study the function of the dopamine system [36]. For example, in experimentally 1-methyl-4-phenyl-1,2,3,6-tetrahydropyridine (MPTP)-lesioned rhesus monkeys, the BOLD signal was found to be positively correlated with the number of surviving dopaminergic neurons [37]. Importantly, because the response of these neurons to dopaminergic agonists is lesion-dependent, phMRI can be utilized clinically to assess the efficacy of therapies in Parkinson's disease patients [38]. Although fewer phMRI studies have focused on the acetylcholine system [39, 40], it is plausible that they also will yield clinical applications.

Studies Discussed in this Chapter

In the present chapter, we discuss studies in which muscarinic receptor function was assessed in a cohort of cognitively tested elderly female rhesus macaques *in vivo* using phMRI and *in vitro* using saturation binding assays. Based on the number of trials to reach a criterion performance in the food port spatial maze navigational test [32], the monkeys were classified into two groups: good spatial performers (GSP) and poor spatial performers (PSP). In the phMRI experiment, the animals were given scopolamine and BOLD signal changes in the PFC and hippocampus were assessed. Subsequently, postmortem M_1 and M_2 receptor binding assays were performed using the PFC and hippocampus tissues of these animals. Finally, the potential relationships between circadian activity and the percentage change in BOLD signal and number and binding affinity of M_1 and M_2 receptors were explored.

Nonhuman Primate Spatial Maze

A nonhuman primate spatial maze was developed using a cohort of elderly female rhesus macaques [32]. In this maze, a bank of 10 food ports was mounted to the wall (Figure 1). Each port consisted of a one-way opaque swinging door. Freely moving monkeys were placed in the room and were allowed to search the ports. One of the 10 food ports was repeated baited with preferred candy until the animals searched the port on the first or second search, three times in a row (location 1).

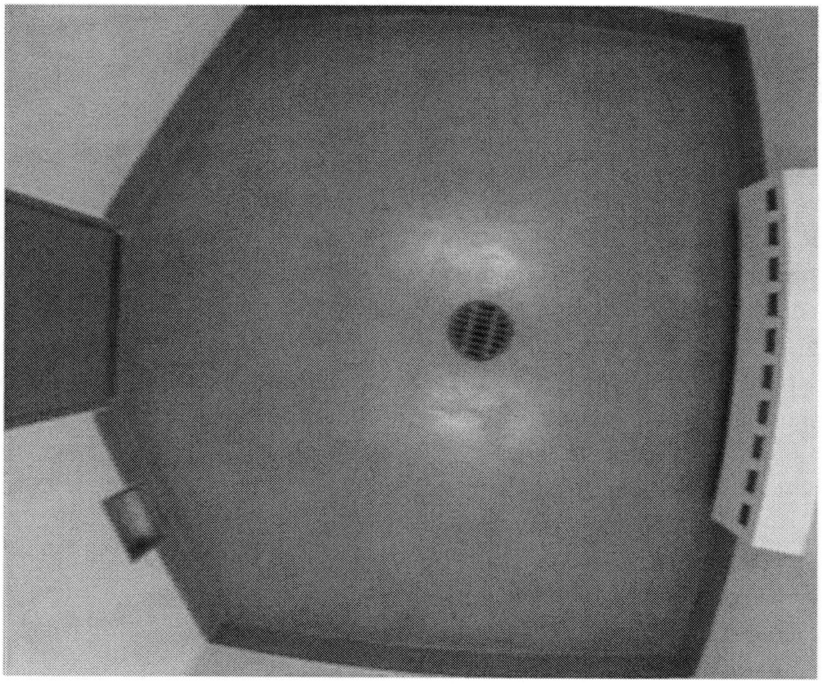

Figure 1. Image of the Spatial Foodport Maze Room. The room is 2.44 x 3.45 m and consists of a human entry door (equipped with a one-way observation window), animal entry door, one-way observation window (not visible in the image), and a custom made 157 x 38 cm bank of 10 ports. The port apparatus is located on the wall opposite the entry doors. Each port measuring 7.6 x 10.2 x 7.6 cm has a one-way opaque swinging door with 5 cm between each port. Cameras were mounted in recesses in the ceiling to record animal behavior in the room and during the task.

Once the monkeys learned the first location, the baited food port was switched to a different one (location 2). Again, the monkeys were allowed to search the series of food ports until they searched the baited port on the first or second port searched three consecutive times.

The number of trials to criterion was counted as well as the search strategies. Monkeys that used a spatial search (searching within 2 ports of the baited port) completed the task in fewer trials than the monkeys that utilized a serial search pattern (searching the ports consecutively).

Circadian Activity

During the testing period, animals wore a nylon activity collar for 2 weeks. Day activity, night activity, number of times the animals woke during the night (wake bouts), and the time to fall asleep (sleep latency), were recorded and averaged over the 2 weeks.

Once the spatial maze was completed, the activity measures were correlated with performance on the spatial maze. Monkeys that performed better on the spatial maze had higher indices of sleep quality than those who performed poorly.

Furthermore, the better performing monkeys were more active during the day than those who performed poorly [32]. After this task was completed, it appeared that there were 2 separate groups of monkeys; one that performed well on the task and one that did not.

BEHAVIORAL OR COGNITIVE CLASSIFICATION

Based on the performance of the cohort in the spatial maze, individual monkeys were separated into two groups: good spatial performers (GSP) and poor spatial performers (PSP). GSP was defined as animals that completed the task in less than 30 trials in either the first or second position (18.7 ± 4.8 trials n = 4 first location; 7 and 14 trials n = 2 for the second location). PSP was defined as animals that completed the task in more than 30 trials (n = 6; 81.8 ± 8.5 trials). Control animals for the phMRI experiment were animals from the same cohort that did not participate in the spatial maze testing (n = 5).

Although the number 30 might seem arbitrary, examining the data, there were monkeys that learned the first location very quickly (less than 30 trials). Or, if the monkeys did not learn the first location quickly, they were able to release the idea of the first location and learned the second location very quickly. This observation is observed in the average number of trials in the first location and the two data points for the second location. Not every monkey exhibited the same pattern of learning. As for the PSP, using 30 trials as the cut-off was not detrimental to those monkeys in the PSP group, as the number of trials to criterion was much higher in the PSP compared to the GSP. There were no monkeys in the PSP group that completed the task in less than 40 trials to criterion.

There were 3 monkeys that were unable to complete the task due to temperament issue or lack of ability to be trained to search the ports. These monkeys would also be considered as PSP, however since there was no clear delineation, they were excluded from the remaining study.

As previously mentioned, during the cognitive testing phase of the experiment, circadian activity was measured in the participating monkeys. The activity was monitored using the Actiwatch system for 2 weeks. Interestingly, we found that the monkeys that performed better on the spatial maze were more active during the day, less active at night, and took less time to fall asleep [32]. Further analysis demonstrated that GSP animals had significantly higher overall activity, higher day activity, and higher Light:Dark activity ratios than PSP animals (Table 1 [41]).

Table 1. Circadian activity levels of good spatial performers (GSP) and poor spatial performers (PSP)[1]

	GSP	PSP
Overall Activity	117.8 ± 13.0*	89.0 ± 5.8
Day Activity	213.0 ± 24.2*	143.2 ± 17.7
Night Activity	24.3 ± 2.3	22.3 ± 2.6
Light:Dark Activity Ratio	8.7 ± 0.8**	6.2 ± 0.5
Sleep Latency	21.8 ± 4.7	33.8 ± 5.1

[1]Units for Overall Activity, Day Activity, and Night Activity are average activity counts. Sleep latency is average time (min) to fall asleep. GSP (n = 6) compared to PSP (n = 6). *$P < 0.05$, **$P < 0.01$ GSP significantly greater than PSP.

PHMRI

Based on their performance in the spatial maze, we hypothesized that a performance difference observed in these monkeys was related to the cholinergic system, specifically the muscarinic receptor system as it has a major role in spatial learning and memory. Furthermore, studies suggest that the function of the acetylcholine system declines with age and is highly affected by Alzheimer's disease. A better understanding of how the muscarinic receptor functions among the different behavioral or cognitive groups of monkeys might yield valuable information on how neurobiology reflects behavior and cognition. To test this hypothesis, the cohort of monkeys from the spatial maze was first exposed to phMRI. In the phMRI, BOLD response to scopolamine, a non-specific muscarinic receptor antagonist, was measured. Monkeys were fasted the night before and the morning of MRI experiments. Monkeys were sedated transported from their home cages to the magnet located on the ONPRC campus. Animals were immediately intubated and respirated with isoflurane, and a MRI compatible catheter was inserted into the cephalic vein. The animal's head was placed inside an MRI coil and fitted with pads to eliminate head movement. A baseline scan of 20 contiguous T_2*-weighted multi-shot echo planar images were obtained over the course of 7 minutes (Figure 2). Immediately following the baseline scans, scopolamine (0.4 mg in 1 ml of isotonic saline or 1 ml of isotonic saline for control animals) was infused over 1 minute using an infusion pump. Following the intravenous infusion, a T_2-weighted and 2 T_1-weighted anatomical images were acquired, which was a total of 46 minutes, sufficient time for scopolamine to have CNS effects [42]. Immediately following the anatomical scans, a second BOLD series of T_2*-weighted images was acquired using the exact parameters as the baseline scan ("challenge scan"). The change in BOLD signal from baseline to challenge scan (scopolamine or vehicle) was determined. Isotonic saline control was used to ensure that the change observed between the baseline and challenge scan was due to a scopolamine effect, and not an artifact from signal drift of the magnet.

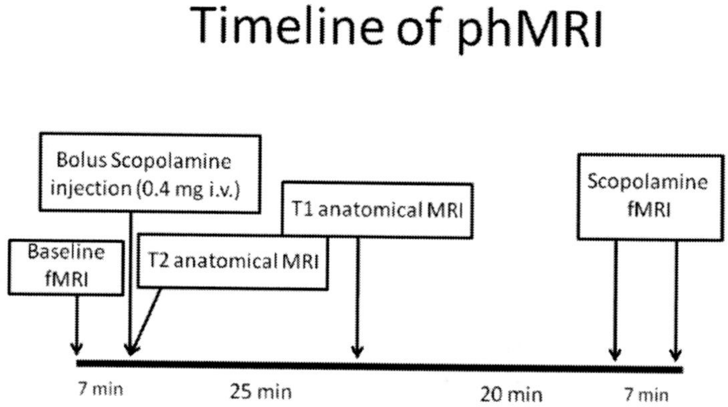

Figure 2. Timeline for pharmacological MRI (phMRI). Once animals were anesthetized, intubated, connected to physiological monitors, and placed in the scanner, a baseline multi-shot T_2* image series was collected. Immediately following the conclusion of the baseline scan, scopolamine was intravenously infused. During the 45 minute time period for the scopolamine to have action within the brain, anatomical images, both T_2 and two T_1 images, were acquired. At the conclusion of the anatomical scans, another scan with the exact parameters of the baseline scan was conducted and defined as a "challenge scan".

Slice-dependent time shifts were interpolated, odd-even slice intensity differences removed, motion corrected, and T_1 and T_2 anatomical scans were co-registered. Baseline and challenge phMRI scans were averaged across the 20 volumes and the single image was normalized to the co-registered anatomical scan of each individual animal (PFC: Figure 3; hippocampus: Figure 4). Regions of Interest (ROI) were drawn for the PFC and hippocampus based on the anatomical scan using key neuroanatomical landmarks from anatomical scans. The ROI was drawn for each animal and was the same ROI for the baseline and scopolamine scan, both of which were co-registered to the anatomical scans.

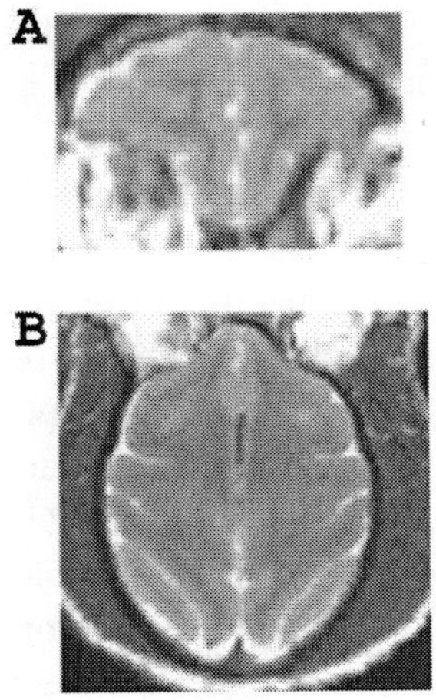

Figure 3. Representative structural MRI images from the prefrontal cortex (PFC) in A.coronal view and B.horizontal views. Regions of interest (ROIs) were drawn on structural images, using a functional overlay, to identify the areas of interest. Voxel intensity of the ROI in the GSP and PSP groups was compared between the baseline and challenge scans. No difference between GSP and PSP in voxel intensity was observed in the PFC.

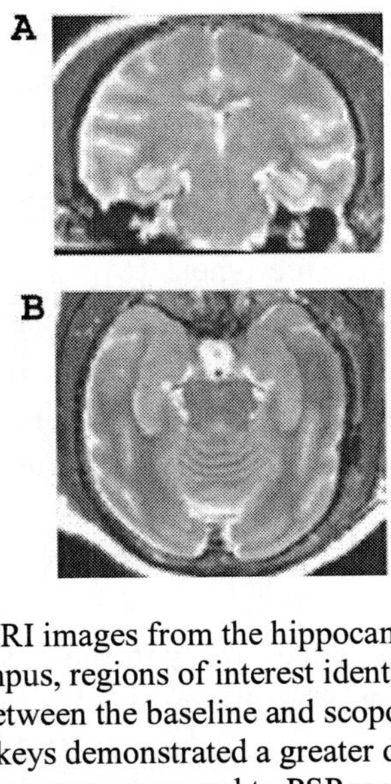

Figure 4. Representative structural MRI images from the hippocampus in A. coronal and B.horizontal views. Like the PFC, in the hippocampus, regions of interest identified the hippocampus, using a functional overlay. Voxel intensity between the baseline and scopolamine challenge scans were compared for each animal. GSP monkeys demonstrated a greater difference in voxel intensity between the baseline and scopolamine challenge scan compared to PSP monkeys.

In the phMRI, control animals administered isotonic saline did not show a significant change in BOLD signal in the PFC or hippocampus. In the hippocampus, but not the PFC, the baseline BOLD signal was significantly lower in the GSP compared to the PSP group. Following scopolamine administration, the BOLD signal change was significantly greater in the GSP than the PSP group in the hippocampus but not the PFC. The phMRI results suggest that there is a different neuronal connectivity of the muscarinic receptor system in the hippocampus compared to the PFC, based on the baseline BOLD signal, and that the muscarinic receptor system in the hippocampus was more functional in the GSP than the PSP group.

Pharmacological intervention and anesthesia are important variables to consider with phMRI in animals. We describe an experiment with scopolamine, a nonspecific muscarinic receptor antagonist. The half life of scopolamine is 17 min in plasma [43]. However, it crosses the blood brain barrier quickly [44] and the effects in the CNS are present for up to 8 hours [45]. Peripherally administered scopolamine binds to muscarinic receptors within the brain [43, 46] and alters EEG activity in healthy, resting humans [47]. In order to anesthetize the animals, we used isoflurane and were able to measure significant BOLD signal changes. This is consistent with other studies, demonstrating that the brain responds to stimuli during anesthesia, even at high doses [48, 49] and that isoflurane does not fully suppress EEG activity [50-52]. Experimental fMRI studies demonstrate that isoflurane anesthesia between the range of 0.9% and 1.25% preserves a measurable BOLD signal [53] level in the anesthetized monkey brain [53-55]. Isoflurane was appropriate for this phMRI study because its effects are not mediated through the cholinergic system nor does it suppress acetylcholine activity or receptor function [56, 57].

Receptor Binding Assays

To validate the phMRI results and confirm our interpretation of the results, we performed post-mortem tissue analysis on the GSP and PSP tissue (Figure 5). For this particular data set, age-related changes in muscarinic receptor function were assessed in rhesus macaques in brain areas important for spatial learning and memory, the PFC, hippocampus, and temporal cortex. The age ranges for the monkeys were adult (5-9 years), middle-aged (11-16 years), and aged (22-30 years) male and female rhesus macaques (*Macaca mulatta*). Age-related changes were observed in the muscarinic type 1 (M_1) and type 2 (M_2) receptors in the PFC, hippocampus (Figure 6. A.) and temporal cortex. Maximum binding of M_1 receptor decreased with age, however there was no change in M_2 receptors. M_1 receptor binding affinity did not significantly change with age, yet the M_2 receptor binding affinity significantly decreased in the PFC and hippocampus [41]. Therefore, the muscarinic receptor system is differentially affected by age.

The data from the age-related changes in muscarinic binding suggest that the changes follow differing timelines in different brain regions and are not universal. Other papers have reported an age-related change in M_1 receptors in the elderly, yet these studies are slightly different from our results. The potential discrepancy in these studies is the age range used within each study. In the study described here, we used adult, middle-aged and aged animals, and our aged animals were 22-30 years. In contrast, previous groups have only used animals that were only 20 years old [24] or did not include middle-aged animals [25]. Therefore, with the increased age range, we identified more subtle age-related changes, not only from adult to middle-age, but also those that occur in the oldest-old. The middle-age and oldest-old age groups are important groups as they appear to be more reflective of the current aging human population and increase the translational properties of the study. The age-related changes in the acetylcholine system have been well documented in animal models as well as humans [13, 58] and are suggested to have an integral role in some of the neuropathology associated with Alzheimer's disease [29, 59-61]. The increase in M_1 receptors in the aged population could possibly be part of a compensatory mechanism in response to age-related decreases in acetylcholine [58, 62]. M_1, and to a lesser extent M_2, are highly expressed in areas of the brain associated with cognitive function and are hypothesized to have a role in cognitive function. Therefore, it is conceivable that elderly cognitively tested animals have a different pattern of M_1 and M_2 expression, dependent on their cognitive performance.

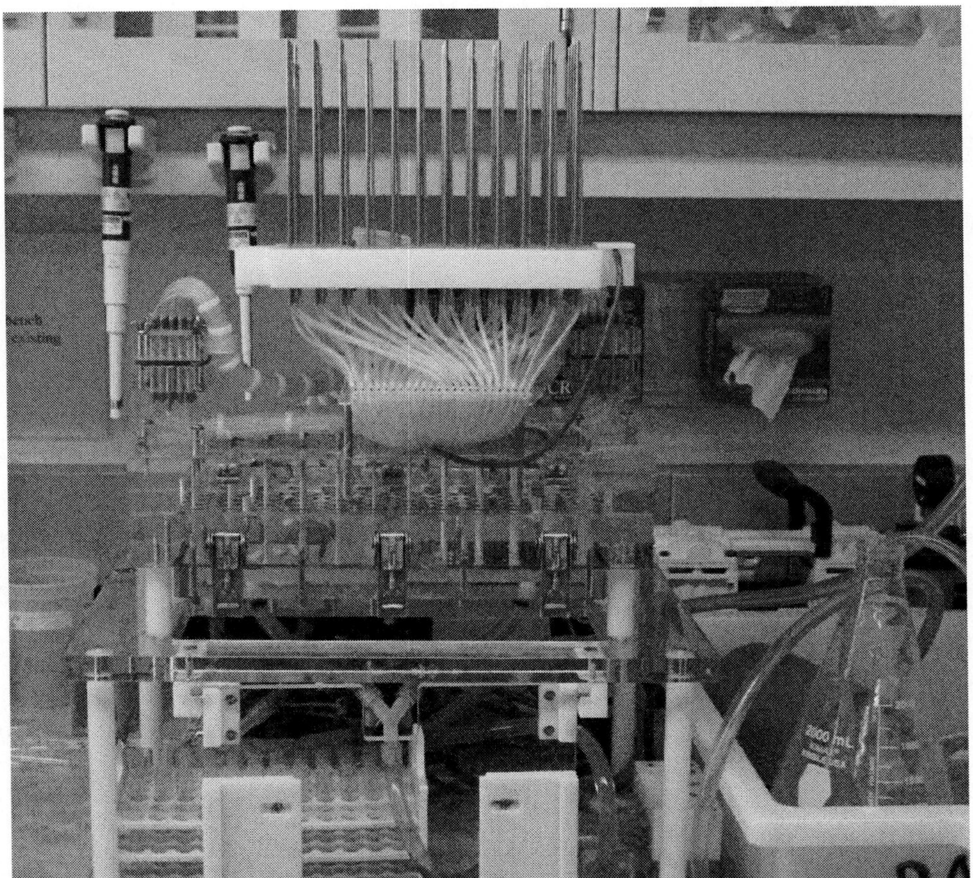

Figure 5. Receptor binding assay set-up. Receptor binding technique uses a cell harvester, pictured here. Radioligands are mixed with tissue and incubated for 1 hour. The mixture is then separated using a cell harvester and radioligands bound to receptors are collected on filter paper. Scintillation fluid is added to the filter paper and the amount of radioligand present is counted in a Beta-counter to determine amount of bound radioligand present in the sample.

To determine if there were individual differences in the M_1 and M_2 maximum receptor binding and receptor affinity potentially reflective of spatial performance, we performed M_1 and M_2 receptor binding assays on PFC and hippocampus tissue from the cognitively tested monkeys. Receptor binding assays on the cognitively tested monkeys revealed a significantly higher maximum binding of M_1 in the hippocampus of GSP compared to PSP (Figure 6.B.). However, no difference in M_1 maximum binding was observed between GSP and PSP in the PFC. M_1 receptor affinity in the hippocampus was significantly higher in the GSP compared to the PSP group, yet no difference was observed between the groups in the PFC. Contrary to the M_1 difference, no difference in M_2 maximum binding or receptor affinity was observed in the PFC (Figure 7. A.). The receptor binding data support our hypothesis that the scopolamine phMRI is reflective of the neurobiology.

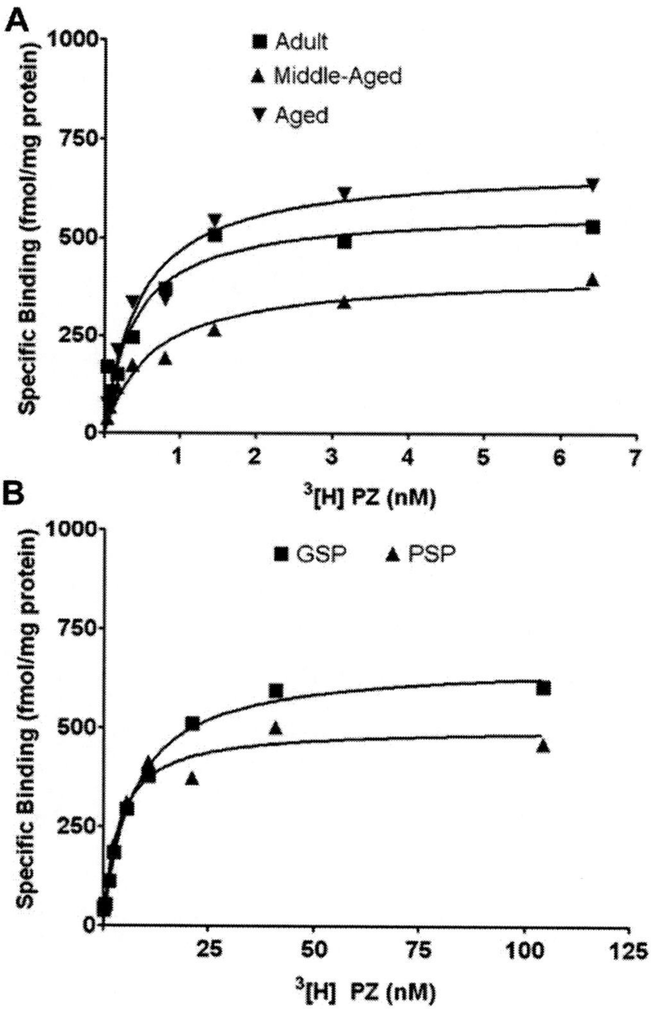

Figure 6. Receptor binding curves for muscarinic type 1 (M_1) receptor in the hippocampus. The x-axis is the number x1000 for both graphs. A. Receptor binding curves of adult (n = 5), middle-aged (n = 5), and aged monkeys (n = 5). B. Receptor binding curves of old female monkeys who were classified as GSP (n = 6) or PSP (n = 6). Age related changes are observed in the M_1 receptor binding number but not affinity. The GSP group had significantly higher receptor number and receptor affinity compared to the PSP group.

Comparing the results from the phMRI and receptor binding results, there is a pattern of increased BOLD signal changes, higher M_1 maximum binding, and stronger M_1 affinity, in the hippocampus but not the PFC. This pattern suggests that the activity of the neurons in the hippocampus of the cognitively tested female animals has a greater reliance on muscarinic receptors than the neurons of the PFC. Alternatively, it is possible the difference observed between the hippocampus and PFC is a result of experience with a hippocampus dependent task. Following scopolamine challenge, there was a significantly greater increase in BOLD signal in the hippocampus of the GSP than PSP group, which appears to be related to cognitive performance and not age because no age differences were observed between the groups [41]. A similar pattern of the increase in BOLD signal change was observed in the PFC but it did not reach statistical significance. A decrease in cognitive performance is observed following scopolamine injection [40, 63, 64] creating a seemingly counterintuitive increase in BOLD signal. However, scopolamine is an antagonist and antagonism of the muscarinic receptor system could initiate downstream effects on neurotransmitters, including acetylcholine, to be released locally to compensate [65, 66], thereby increasing the BOLD signal (Figure 7.B). The neurotransmitter system could in turn initiate the recruitment of other neuronal populations, increasing neuronal activity, which would be measured as an increase in BOLD signal. Scopolamine also targets other muscarinic receptors, including presynaptic muscarinic receptors, and blockade of those receptors can initiate the local release of acetylcholine, which is measured within the brain with microdialysis [65]. Thus it is feasible that a more specific antagonist for the M_1 receptor might have an alternate BOLD effect than scopolamine. Alternatively, scopolamine injection could saturate muscarinic receptors in the PSP hippocampus, but was not a sufficient dose to saturate the muscarinic receptors in the GSP hippocampus; this might account for the difference between these two groups.

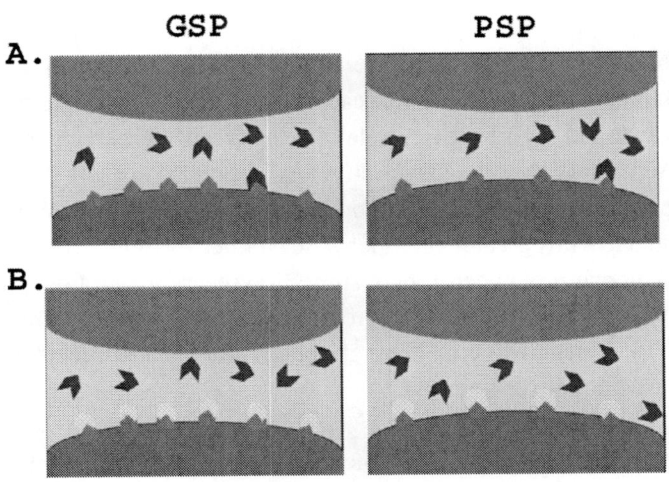

Figure 7. Schematic drawing of the hypothesized action of muscarinic receptor function in the hippocampus, based on phMRI and receptor binding assays, in good spatial performers (GSP) compared to poor spatial performers (PSP). A. In the phMRI, GSP had a greater BOLD response to scopolamine compared to PSP. At the synapse (pre-synaptic terminals in orange; post-synaptic neuron in blue), scopolamine (yellow) binds to muscarinic type 1 (M_1) receptors (green), increasing the BOLD response, potentially through increasing local acetylcholine (red). B. Receptor binding assays corroborated the phMRI indicating that the GSP had a significantly greater number of M_1 (green) which had a greater affinity for the ligand acetylcholine (red) compared to PSP.

Based on these data, our hypothesis is that the higher number of muscarinic receptors in the GSP than PSP group creates a greater compensatory effect, resulting in a greater increase in BOLD signal following a scopolamine challenge. Therefore, the BOLD signal following scopolamine might indicate the plasticity of the brain to compensate for a temporary blockade of the muscarinic receptor system and the brains showing more plasticity might be those which demonstrate better cognitive performance. The receptor binding assays are consistent with this hypothesis and corroborate the phMRI findings. Moreover, the receptor binding assays appear to indicate that the M_1 receptor might have a greater role in the scopolamine blockade than M_2 within the PFC and the hippocampus. In the hippocampus, the number of M_1 receptors and the receptor binding affinity is higher in the GSP than PSP group. Thus, a greater increase in BOLD signal in the hippocampus is associated with a greater number and binding affinity of M_1 receptors. The M_2 receptor number and affinity was not different between the two groups, suggesting that M_2 have a minor role in spatial learning and memory [23]. In the PFC, there was more variability in the baseline BOLD signal as well as the percent change of BOLD signal compared to the hippocampus. As the groups were defined by performance in the spatial maze, this anatomical difference might be related to the fact that spatial learning and memory heavily relies on the hippocampus [67-70].

Interestingly, there was lower baseline BOLD signal in the hippocampus of the GSP than the PSP group. The lower baseline BOLD signal could have enabled a greater response measured by scopolamine injection in the GSP solely due to starting at a lower level. Conversely, the lower baseline BOLD signal in the hippocampus of the GSP group might indicate a different neuronal connectivity [71], observed when functional connectivity MRI studies are performed [72, 73], which could contribute to their enhanced cognitive performance in the spatial foodport maze [32]. Circadian activity was correlated with performance in the spatial food port maze [32].

Animals that had better sleep quality indices performed better on the spatial maze (Table 1). GSP animals had significantly higher overall activity, day activity, and Light:Dark activity ratios than PSP animals. Comparing the circadian activity data with the phMRI and receptor binding results, we found significant correlations between the percentage change in BOLD signal and Day Activity (Table 2). Day Activity also correlated with M_1 maximum binding and receptor affinity in the hippocampus. Both correlations suggest that more Day Activity is observed in animals with greater number and higher affinity. These data support an interconnected relationship between hippocampal M_1 receptor function, spatial learning and memory, and circadian activity. Therefore, it is possible that the negative circadian effect on cognition is due to a negative circadian effect on muscarinic receptor function, in turn resulting in decreased cognitive function.

Table 2. Correlations between day time activity and percentage change in BOLD following a scopolamine challenge, M_1 maximum receptor binding and M_1 receptor affinity

Correlation	r value
Day Activity-% Change in BOLD	0.70**
Day Activity-M1 Bmax	-0.54*
Day Activity-M1 1/Kd	0.65*

*$P < 0.05$, **$P < 0.01$. A total of 12 animals were used for each correlation

CONCLUSION

Collectively, the data from these studies demonstrate that 1) freely moving monkeys can learn a spatial maze, 2) spatial maze performance is associated with circadian activity in elderly female rhesus monkeys, 3) M_1 receptor function is associated with spatial performance and circadian activity, and 4) scopolamine phMRI can be used for assessment of muscarinic receptor function in the context of cognitive performance and circadian activity. The studies presented here demonstrate that grouping based on performance is valuable in identifying neurobiological differences between the groups. As behavioral and cognitive performance might be reflective of the assessed neurobiological measures, using behavioral categorizations can yield valuable data. For the presented data, it is possible that various treatments might not yield circadian consequences but could negatively affect behavioral or cognitive functions. It is also possible that a treatment could negatively affect M_1 receptor function and, in turn, negatively impact spatial learning and memory, yet it may not alter circadian activity due to compensatory mechanisms. Increased efforts are warranted to use behavioral categorization for translational research of neuropsychological and neurological disorders in humans to identify their underlying neurobiological consequences.

ACKNOWLEDGMENTS

This work was supported by NIH grants RR-000163, AG-023477, AG-029612, AG-019100, AG-024978 (OHSU Roybal Center), AG-026472, and AG-027697 and an OHSU Tartar Fellowship.

REFERENCES

[1] Haley, G.E., et al., Age-related decreases in SYN levels associated with increases in MAP-2, apoE, and GFAP levels in the rhesus macaque prefrontal cortex and hippocampus. *Age* (Dordr), 2010. 32(3): p. 283-96.

[2] Haley, G.E., et al., Effects of epsilon4 on Object Recognition in the Non-Demented Elderly. *Curr. Aging Sci*, 2010. 3(2): p. 127-137.

[3] Haley, G.E. and F.W. Flynn, Agonist and hypertonic saline-induced trafficking of the NK3-receptors on vasopressin neurons within the paraventricular nucleus of the hypothalamus. *Am. J. Physiol. Regul. Integr. Comp. Physiol*, 2006. 290(5): p. R1242-50.

[4] Haley, G.E. and F.W. Flynn, Blockade of NK3R signaling in the PVN decreases vasopressin and oxytocin release and c-Fos expression in the magnocellular neurons in response to hypotension. *Am. J. Physiol. Regul. Integr. Comp. Physiol*, 2008. 295(4): p. R1158-67.

[5] Thathiah, A. and B. De Strooper, G protein-coupled receptors, cholinergic dysfunction, and Abeta toxicity in Alzheimer's disease. *Sci. Signal*, 2009. 2(93): p. re8.

[6] Nathanson, N.M., Synthesis, trafficking, and localization of muscarinic acetylcholine receptors. *Pharmacology and Therapeutics*, 2008. 119(1): p. 33-43.

[7] Wisman, L.A., et al., Functional convergence of dopaminergic and cholinergic input is critical for hippocampus-dependent working memory. *J. Neurosci*, 2008. 28(31): p. 7797-807.
[8] Gage, F.H., et al., Experimental approaches to age-related cognitive impairments. *Neurobiol. Aging*, 1988. 9(5-6): p. 645-655.
[9] Fredrickson, A., et al., The use of effect sizes to characterize the nature of cognitive change in psychopharmacological studies: an example with scopolamine. *Human Psychopharmacologia*, 2008. 23(5): p. 425-436.
[10] Thomas, E., et al., Specific impairments in visuospatial working and short-term memory following low-dose scopolamine challenge in healthy older adults. *Neuropsychologia*, 2008. 46(10): p. 2476-2484.
[11] Elrod, K. and J.J. Buccafusco, An evaluation of the mechanism of scopolamine-induced impairment in two passive avoidance protocols. *Pharmacol. Biochem. Behav*, 1988. 29(1): p. 15-21.
[12] Biggan, S.L., J.L. Ingles, and R.J. Beninger, Scopolamine differentially affects memory of 8- and 16-month-old rats in the double Y-maze. *Neurobiol. Aging*, 1996. 17(1): p. 25-30.
[13] Araujo, J.A., C.M. Studzinski, and N.W. Milgram, Further evidence for the cholinergic hypothesis of aging and dementia from the canine model of aging. *Prog. Neuropsychopharmacol. Biol. Psychiatry*, 2005. 29(3): p. 411-22.
[14] Savage, U.C., et al., Effects of scopolamine on learning and memory in monkeys. *Psychopharmacology* (Berl), 1996. 123(1): p. 9-14.
[15] Taffe, M.A., M.R. Weed, and L.H. Gold, Scopolamine alters rhesus monkey performance on a novel neuropsychological test battery. *Brain Res. Cogn. Brain Res*, 1999. 8(3): p. 203-12.
[16] Rosier, A., L. Cornette, and G.A. Orban, Scopolamine-induced impairment of delayed recognition of abstract visual shapes. *Neuropsychobiology*, 1998. 37(2): p. 98-103.
[17] Flynn, D.D., et al., Differential alterations in muscarinic receptor subtypes in Alzheimer's disease: implications for cholinergic-based therapies. *Life Sci.*, 1995. 56(11-12): p. 869-76.
[18] Flynn, D.D., et al., Differential regulation of molecular subtypes of muscarinic receptors in Alzheimer's disease. *J. Neurochem*, 1995. 64(4): p. 1888-91.
[19] Tamminga, C.A., The neurobiology of cognition in schizophrenia. *J. Clin. Psychiatry*, 2006. 67(9): p. e11.
[20] Rouse, S.T., et al., Localization of M(2) muscarinic acetylcholine receptor protein in cholinergic and non-cholinergic terminals in rat hippocampus. *Neuroscience Lett.*, 2000. 284(3): p. 182-6.
[21] Jagoda, E.M., et al., Regional brain uptake of the muscarinic ligand, [18F]FP-TZTP, is greatly decreased in M2 receptor knockout mice but not in M1, M3 and M4 receptor knockout mice. *Neuropharmacology*, 2003. 44(5): p. 653-61.
[22] Gautam, D., et al., M1-M3 muscarinic acetylcholine receptor-deficient mice: novel phenotypes. *J. Mol. Neurosci*, 2006. 30(1-2): p. 157-60.
[23] Matsui, M., et al., Functional analysis of muscarinic acetylcholine receptors using knockout mice. *Life Sci*, 2004. 75(25): p. 2971-81.

[24] Wagster, M.V., et al., Laminar Organization and Age-Related Loss of Cholinergic Receptors in Temporal Neocortex of Rhesus Monkey. *J. Neurosci*, 1990. 10(9): p. 2879-2885.

[25] Vannucchi, M.G. and P.S. Goldman-Rakic, Age-dependent decrease in the affinity of muscarinic M1 receptors in neocortex of rhesus monkeys. *Proc. Natl. Acad. Sci. U S A*, 1991. 88(24): p. 11475-9.

[26] Ma, L., et al., Selective activation of the M1 muscarinic acetylcholine receptor achieved by allosteric potentiation. *Proc. Natl. Acad. Sci. U S A*, 2009. 106(37): p. 15950-5.

[27] Fisher, A., et al., M1 muscarinic agonists can modulate some of the hallmarks in Alzheimer's disease: implications in future therapy. *J. Mol. Neurosci*, 2003. 20(3): p. 349-56.

[28] Van Kampen, J.M. and C.B. Eckman, Agonist-induced restoration of hippocampal neurogenesis and cognitive improvement in a model of cholinergic denervation. *Neuropharmacology*, 2009. In Press.

[29] Fisher, A., Cholinergic treatments with emphasis on m1 muscarinic agonists as potential disease-modifying agents for Alzheimer's disease. *Neurotherapeutics*, 2008. 5(3): p. 433-42.

[30] Haimov, I., E. Hanuka, and Y. Horowitz, Chronic insomnia and cognitive functioning among older adults. *Behav. Sleep Med*, 2008. 6(1): p. 32-54.

[31] Morin, C.M., F.C. Blais, and V. Mimeault, Sleep disturbances in late life, in *Handbook of clinical geropsychology*, H. M. and V.B. Van Hasselt, Editors. 1998, Plenum Press: New York. p. 273-299.

[32] Haley, G.E., et al., Circadian activity associated with spatial learning and memory in aging rhesus monkeys. *Exper. Neurol*, 2009. 217(1): p. 55-62.

[33] Gillette, M.U., et al., Role of the M1 receptor in regulating circadian rhythms. *Life Sci.*, 2001. 68(22-23): p. 2467-72.

[34] Liu, C. and M.U. Gillette, Cholinergic regulation of the suprachiasmatic nucleus circadian rhythm via a muscarinic mechanism at night. *J. Neurosci.*, 1996. 16(2): p. 744-51.

[35] Wise, R.G. and I. Tracey, The role of fMRI in drug discovery. *J Magn. Reson. Imaging*, 2006. 23(6): p. 862-76.

[36] Knutson, B. and S.E. Gibbs, Linking nucleus accumbens dopamine and blood oxygenation. *Psychopharmacology* (Berl), 2007. 191(3): p. 813-22.

[37] Zhang, Z., et al., Assessing nigrostriatal dysfunctions by pharmacological MRI in parkinsonian rhesus macaques. *Neuroimage*, 2006. 33(2): p. 636-43.

[38] Luan, L., et al., Pharmacological MRI (phMRI) monitoring of treatment in hemiparkinsonian rhesus monkeys. *Cell Transplant*, 2008. 17(4): p. 417-25.

[39] Wink, A.M., et al., Age and cholinergic effects on hemodynamics and functional coherence of human hippocampus. *Neurobiol. Aging*, 2006. 27: p. 1395-1404.

[40] Thiel, C.M., Cholinergic modulation of learning and memory in the human brain as detected with functional neuroimaging. *Neurobiol. Learning Memory*, 2003. 80: p. 234-244.

[41] Haley, G.E., et al., Hippocampal M1 Receptor Function Associated with Spatial Learning and Memory in Aged Female Rhesus Macaques AGE, 2010. In Press.

[42] Ali-Melkkila, T., J. Kanto, and E. Iisalo, Pharmacokinetics and related pharmacodynamics of anticholinergic drugs. *Acta Anaesthesiol. Scand.*, 1993. 37(7): p. 633-642.
[43] Frey, K.A., et al., Quantitative in vivo receptor binding: I. Theory and application to the muscarinic cholinergic receptor. *J. Neurosci.*, 1985. 5(2): p. 421-428.
[44] Frey, K.A., B. Ciliax, and B.W. Agranoff, Quantitative in vivo receptor binding. IV: Detection of muscarinic receptor down-regualtion by equilibrium and by tracer kinetic methods. *Neurochem. Res.*, 1991. 16(9): p. 1017-1023.
[45] Ali-Melkkila, T., J. Kanto, and E. Iisalo, Pharmacokineticss and related pharmacodynapics of anticholinergic drugs. *Acta Anaesthesiol. Scand.* 1993. 37(7): p. 633-642.
[46] Wamsley, J.K., et al., Muscarinic cholinergic receptors: autoradiographic localization of high and low affinity agonist binding sites. *Brain Res.*, 1980. 200: p. 1-12.
[47] Ebert, U., et al., Pharmacokinetic-pharmacodynapic modeling of the electroencephalogram effects of scopolamine in healthy volunteers. *J. Clin. Pharmacol.*, 2001. 41: p. 51-60.
[48] Porkkala, T., et al., Median nerve somatosenory evoked potentials during isoflurane anaesthesia. *Can. J. Anaesthesia*, 1997. 44: p. 963-968.
[49] Hartikainen, K. and M. Rorarius, Cortical responses to auditory stiuli during isoflurane burst suppression. *Anaesthesia*, 1999. 54: p. 210-214.
[50] Tsushima, K., et al., Supressive actions of volatile anaesthetics on the response capability in cats. *Can. J. Anaesthesia*, 1998. 45: p. 240-245.
[51] Ogawa, T., et al., The divergent actions of volatile anaesthetics on background neuronal activity and reactive capability in the central nervous system in cats. *Can. J. Anaesthesi*a, 1992. 39: p. 862-872.
[52] Murrell, J.C., D. Waters, and C.B. Johnson, Comparative effects of halothane, isoflurane, sevoflurane, and desflurane on the electroencephalogram of the rat. *Laboratory Animals*, 2008. 42: p. 161-170.
[53] Vincent, J.L., et al., Intrinsic functional architecture in the anaesthetized monkey brain. *Nature*, 2007. 447: p. 83-86.
[54] Logothetis, N.K., et al., Functional imaging of the monkey brain. *Nature Neurosci.*, 1999. 2(6): p. 555-562.
[55] Small, S.A., et al., Imaging correlates of brain function in monkeys and rats isolates a hippocampal subregion differentially vulnerable to aging. *Proc. Natl. Acad. Sci. U S A*, 2004. 101(18): p. 7181-7186.
[56] Eger, E.I., et al., Acetylcholine receptors do not mediate the immobilization prodcued by inhaled anesthetics. *Anaesthesia and Analgesia*, 2002. 94: p. 1500-1504.
[57] Nakayama, T., et al., Differential effects of volatile anesthetics on m_3 muscarinic receptor coupling to the $G\alpha_q$ heterotrimeric G protein. *Anesthesiology*, 2006. 105: p. 313-324.
[58] Sarter, M. and J. Turchi, Age- and dementia-associated impairments in divided attention: psychological constructs, animal models, and underlying neuronal mechanisms. *Dement. Geriatr. Cogn. Disord.*, 2002. 13(1): p. 46-58.
[59] Shiozaki, K., et al., Distribution of m1 muscarinic acetylcholine receptors in the hippocampus of patients with Alzheimer's disease and dementia with Lewy bodies-an immunohistochemical study. *J. Neurol. Sci*, 2001. 193(1): p. 23-8.

[60] Pakaski, M. and J. Kalman, Interactions between the amyloid and cholinergic mechanisms in Alzheimer's disease. *Neurochem. Int*, 2008. 53(5): p. 103-11.
[61] Lai, M.K., et al., Selective effects of the APOE epsilon4 allele on presynaptic cholinergic markers in the neocortex of Alzheimer's disease. *Neurobiol. Dis*, 2006. 22(3): p. 555-61.
[62] Costall, B., et al., Biochemical models for cognition enhancers. *Pharmacopsychiatry*, 1990. 23 Suppl 2: p. 85-8; discussion 89.
[63] Klinkenberg, I. and A. Blokland, The validity of scopolamine as a pharmacological model for cognitive impairment: A review of animal behavioral studies. *Neurosci. Biobehav. Rev.*, 2010. In press.
[64] Ebert, U. and W. Kirch, Scopolamine model of dementia: electroencephalogram findings and cognitive performance. *Eur. J. Clin. Invest*, 1998. 28(11): p. 944-9.
[65] Day, J., G. Damsma, and H.C. Fibiger, Cholinergic activity in the rat hippocampus, cortex and striatum correlates with locomotor activity: an in vivo microdialysis study. *Pharmacol. Biochem. Behav.*, 1991. 38(4): p. 723-9.
[66] Scali, C., et al., Peripherally injected scopolamine differentially modulates acetylcholine release in vivo in the young and aged rats. *Neurosci. Lett.*, 1995. 197(3): p. 171-4.
[67] Raber, J., et al., Radiation-induced cognitive impairments are associated with changes in indicators of hippocampal neurogenesis. *Radiat. Res.*, 2004. 162: p. 39-47.
[68] Cho, Y.H., E. Friedman, and A.J. Silva, Ibotenate lesion of the hippocampus impair spatial learning but not contextual fear conditioning. *Behav. Brain Res.*, 1999. 98: p. 77-87.
[69] Banta Lavenex, P. and P. Lavenex, Spatial memory and the monkey hippocampus: not all space is created equal. *Hippocampus*, 2009. 19(1): p. 8-19.
[70] Lavenex, P.B., D.G. Amaral, and P. Lavenex, Hippocampal lesion prevents spatial relational learning in adult macaque monkeys. *J. Neurosci.*, 2006. 26(17): p. 4546-58.
[71] Fransson, P., Spontaneous low-frequency BOLD signal fluctuations: An fMRI investigation of the resting-state default mode of brain function hypothesis. *Human Brain Mapp*ing, 2004. 26(1): p. 15-29.
[72] Qiu, A., et al., Hippocampal-cortical structural connectivity disruptions in schizophrenia: An integrated perspective from hippocampal shape, cortical thickness, and integrity of white matter bundles. *Neuroimage*, 2010. In Press.
[73] Pereira, F.R., et al., Asymmetrical hippocampal connectivity in mesial temporal lobe epilepsy: evidence from resting state fMRI. *BMC Neurosci.*, 2010. 11: p. 66.

In: Monkeys: Biology, Behavior and Disorders
Editor: Rachel M. Williams, pp 137-149
ISBN: 978-1-61209-911-8

Chapter 6

PLANNING ABILITIES OF MONKEYS

***Damian Scarf*[1], *Herbert S. Terrace*[2] *and Michael Colombo*[1,*]**
[1]Department of Psychology, University of Otago, Dunedin, New Zealand
[2]Department of Psychology, Columbia University, New York, NY 10027, USA

As adults we plan things we will do today, tomorrow, next week, next month, and even years from now. Our ability to plan is adaptive in the sense that it allows us to anticipate events that may occur in the future and appropriately prepare for them in the present. For example, it is common for young adults to save for their retirement, an event that may be 30 to 40 years in their future. Although children do not plan on time scales as large as this, they still quickly learn to save their pocket money to buy the latest toy or plan their activities for the coming weekend.

A number of investigators have examined the planning abilities of a range of different nonhuman animals such as chimpanzees (Biro and Matsuzawa, 1999; Mulcahy and Call, 2006), monkeys (Ohshiba, 1997; Paxton and Hampton, 2009), and birds (Scarf and Colombo, 2010; Raby, Alexis, Dickinson, and Clayton, 2007). Among the many different tasks used, the *simultaneous chaining paradigm* has emerged as one of the favourites for studying planning abilities (Terrace, 1984). The task is relatively straightforward and requires the subjects to respond to *n* number of simultaneously presented stimuli in a prescribed order. For ease of exposition we refer to the stimuli with the letters of the alphabet, and the order in which they should be pressed with arrows. For example, the correct order of responding for a five-item list would be A → B → C → D → E. Any deviation from the prescribed order, such as by skipping over an item (e.g., A → B → D), or by responding to an item already pressed (e.g., A → B → C → B) is considered an error.

In the first study to investigate planning using the simultaneous chaining paradigm, the chimpanzee Ai was trained to respond to three Arabic numerals, drawn from the range 0 to 9, in ascending order (Biro and Matsuzawa, 1999). In addition to these 'normal' trials, 'switch' probe trials were introduced. On switch trials, after Ai responded to the first item (A), the positions of the second (B) and third (C) items were exchanged. Ai performed very poorly on switch trials compared to normal trials (45% vs 90%) and, when switch trials were responded

[*]Phone/Fax: 64-3-479-7626 (office), 64-3-479-8335 (fax). E-mail: colombo@psy.otago.ac.nz

to correctly, response latencies to the second item were longer than on normal trials (907 msec vs 501 msec). Biro and Matsuzawa (1999) suggested that the poor performance on switch trials reflected Ai executing a response plan formed at the outset of a trial, and that the increased response time on correct trials reflected Ai's updating of that response plan.

If Ai was indeed planning, then the expectation was that her representation would guide her behaviour if the stimuli were visually present or not. Consequently, in their second set of studies, Matsuzawa and colleagues used a 'mask' task in which following a response to the first item the remaining items were covered by opaque white squares (Inoue and Matsuzawa, 2007; Kawai and Matsuzawa, 2000). In order to correctly complete a mask trial the subject had to rely on a representation of each item's spatial position to guide their behaviour, that is, remember where the stimuli were and the order to which they needed to be responded. Ai was able to respond to four masked numerals above chance while her son, Ayumu, was able to respond to eight. Responding to eight masked numerals above chance is a feat that is yet to be achieved by adult humans, who have only been tested with five masked items (Cook and Wilson, 2010; Silberberg and Kearns, 2009). Together, the performance on the switch and mask tasks suggests that chimpanzees are able to plan a series of actions.

More recently, Beran and colleagues have conducted a similar set of experiments with monkeys (Beran, Pate, Washburn, and Rumbaugh, 2004). The monkeys were trained on a five-item list, and on switch trials, the positions of the second and third items were switched. That is, after responding to stimulus A, the positions of the second (B) and third (C) stimuli were switched, while the positions of the fourth (D) and fifth (E) stimuli remained the same. Similar to Ai, the monkeys performed worse on switch trials compared to normal trials (37.5% vs 55%). On mask trials, the monkeys only responded significantly above chance to the first masked item (B), whereas accuracy to remaining masked items (C, D, and E) was no different from chance. At best, Beran et al.'s (2004) data show that monkeys can plan a single response ahead.

The fact that Beran et al.'s (2004) monkeys were only able to plan one step ahead, whereas Matsuzawa and colleagues' chimpanzees were able to plan several steps ahead may indicate a significant difference in the planning abilities of chimpanzees and monkeys. However, before we can conclude that monkeys have limited planning abilities compared to chimpanzees, we must make sure that they have not been procedurally disadvantaged. On this issue, we wondered whether Beran et al.'s (2004) requirement that their monkeys make responses using a joystick (compared to Matsuzawa and colleagues' chimpanzees who responded directly to a touchscreen) could have accounted for the differences in planning abilities between the two species. Although the use of a joystick may seem like a small issue, it may in fact have serious consequences on memory. For example, Beran et al. (2004) note that in general, responses with a joystick take longer to execute than responses to a screen, a possible consequence of which may have been that Beran's monkeys may have been under a greater memory load than Matsuzawa's chimpanzees, and this greater memory load may have adversely affected the monkeys' planning abilities. Alternatively, the poor performance of Beran et al.'s (2004) monkeys may have been due to the stimulus-response spatial discontiguity that arises when a joystick is used. Given that studies with monkeys have shown that performance on cognitive tasks is adversely affected by large degrees of stimulus-response spatial discontiguity (Iwai, Yaginuma, and Mishkin, 1986), this factor alone could have also accounted for the difference in planning abilities between monkeys and chimpanzees.

EXPERIMENT 1: SWITCH AND MASK PERFORMANCE

In our first experiment (see Scarf, Danly, Morgan, Colombo, and Terrace, in press), we explored whether reducing the stimulus-response spatial discontiguity for monkeys, by training them with a touchscreen, would result in performance levels similar to Matsuzawa and colleagues' chimpanzees (Inoue and Matsuzawa, 2007, 2009; Kawai and Matsuzawa, 2000), or at the very least, an improvement over Beran et al.'s (2004) joystick-trained monkeys.

Method

Four adult male rhesus monkeys (*Macaca mulatta*; Benedict, Horatio, Oberon, and Prospero), served as subjects. All had extensive experience at ordering arbitrary and numerical stimuli (e.g. Brannon, Cantlon, and Terrace, 2006; Brannon and Terrace, 2000; Kornell and Terrace, 2007; Subiaul, Cantlon, Holloway, and Terrace, 2004; Terrace, Son, and Brannon, 2003). They were tested in operant chambers, the front wall of which housed a 15 in 3M MicroTouch™ touchscreen and two speakers located on the top right and top left of the touchscreen. Positioned in front of the touchscreen was a transparent Perspex panel that contained 16 equally spaced rectangular cutouts arranged in a 4 x 4 matrix, each cutout being 1.5 in high by 2 in wide. The five stimuli consisted of pictures of a man (A), a tree (B), a parrot (C), a pair of slippers (D), and a glass cube (E). Reinforcers were 190-mg Noyes (Noyes, Lancaster, NH) banana flavoured pellets.

The training protocol was similar to that used by Terrace et al. (2003). In the first phase of training, all five list items were displayed simultaneously with monkeys learning the correct item order through the process of trial and error until they achieved a performance level of one session with at least 60% correct responses. Correctly pressed stimuli remained on, and the chance probability of responding correctly to all five stimuli was 0.08% (.20 x .25 x .25 x .25 x .25). In the second phase of training, as per Inoue and Matsuzawa (2009), correctly pressed stimuli disappeared following a correct response, and the probability of responding correctly to all stimuli was 0.8% (.20 x .25 x .33 x .50 x 1.0). Subjects were trained on the second phase until they reached a criterion of 70% or more correct responses. In both phases, the spatial position of items was varied from trial to trial to ensure subjects did not learn a rote motor sequence.

Once the monkeys had reached criterion on the second phase of training, switch trials were introduced on 6 trials of a 60-trial session. There were three types of switch trial: BC switch trials (stimuli B and C exchanging places), CD switch trials (stimuli C and D exchanging places), and DE switch trials (stimuli D and E exchanging places). The procedure on a switch trial was as follows. Following the onset of the five stimuli, a response to stimulus A resulted in the respective switch items exchanging places. The birds were first tested for 10 sessions with BC switch trials, then 10 sessions with CD switch trials, and finally 10 sessions with DE switch trials. Each ten-session block yielded 540 normal trials and 60 switch trials.

Following 30 sessions of testing with the three different switch trial types, the monkeys were tested with mask trials. Each session was composed of 60 mask trials in which following a response to item A, all of the remaining list items were replaced by white squares. In the first phase of training, only items A and B were presented. When the subject responded to item A, item B was masked. In the second, third, and fourth phases of training, subjects were trained with three (A → B → C), four (A → B → C → D), and finally five items (A → B → C → D → E). In all cases, following a response to item A the remaining stimuli were masked. Transition between phases was dependent on subjects responding at or above 70% correct to each masked item. For example, in order reach criterion with four items, subjects had to respond correctly to items B, C, and D on at least 70% of the trials within a single session.

Results

Benedict, Horatio, Oberon, and Prospero took 240, 1200, 120, and 1020 trials, respectively, to learn the first phase where the stimuli remained illuminated, and transferred rapidly to the second phase of training (range 60-180 trials) where the stimuli disappeared following a correct response. The performance on BC switch trials is shown in the two leftmost bars of each of the four panels of Figure 1. All four subjects performed significantly better on normal trials than on BC switch trials in which the second and third items exchanged places (Benedict: $t(9) = 4.23, p < .05$; Horatio: $t(9) = 12.01, p < .001$; Oberon: $t(9) = 5.10, p < .05$; Prospero: $t(9) = 2.85, p < .05$). In addition to examining percent correct, we also examined latency, as it is often the case that latency is a more sensitive measure than percent correct.

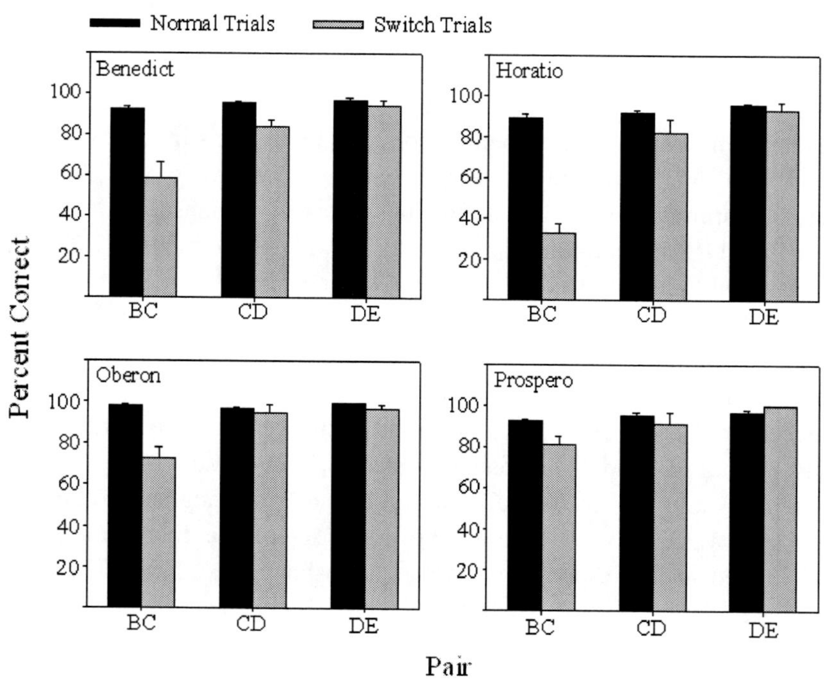

Figure 1. Percent correct on normal and switch trials for the BC switch (two leftmost bars), CD switch (two middle bars), and DE switch (two rightmost bars) for each of the four subjects. Significant differences in performance are indicated with an asterisk.

In line with the findings on percent correct, all subjects took significantly longer to respond to item B on BC switch trials compared to BC normal trials (Benedict: $t(8) = 3.54$, $p < .05$; Horatio: $t(7) = 5.56$, $p < .05$; Oberon: $t(9) = 5.28$, $p < .05$; Prospero: $t(9) = 3.10$, $p < .05$). Together, the poorer performance and longer response latencies suggest that all subjects were planning at least a response to B at the start of a trial.

The performance on CD switch trials is shown in the two middle bars of each of the four panels of Figure 1. In contrast to what we observed for the BC switch trials where all four subjects were impaired, only one subject (Benedict) showed an impairment on the CD switch trials, $t(9) = 3.96$, $p < .05$. With respect to response latencies, two subjects (Benedict and Oberon) took significantly longer to respond to item C on CD switch trials compared to CD normal trials (Benedict: $t(9) = 2.96$, $p < .05$; Oberon: $t(9) = 3.57$, $p < .05$).

The performance on DE switch trials is shown in the two rightmost bars of each of the four panels of Figure 1. Only one subject's performance (Prospero) was significantly different on DE switch trials than on normal trials, $t(9) = 2.77$, $p < .05$, but the direction was opposite to what we would predict if the animal was planning. With respect to response latency, only one subject (Benedict) showed an elevated response latency to respond to item D on DE switch trials compared to DE normal trials, $t(9) = 5.24$, $p < .05$.

The performance of subjects on the mask condition with 3-, 4-, and 5-item list lengths is shown in Table 1. On the 3-item list mask task, in which items B and C are masked, all subjects reached the criterion of 70% or more correct responses to item B ($p < .05$, binomial test). On the 4-item mask task, however, only one subject (Benedict) reached the criterion within the allotted 40 sessions (2400 trials) of training. Horatio, Oberon, and Prospero responded above chance to the first masked item (B), but their performance fell to chance on the third item. Due to the failure of Horatio, Oberon, and Prospero on the 4-item mask task, only Benedict was run on the final 5-item mask task, and he was unable to reach criterion despite 45 sessions (2700 trials) of training.

Table 1. Performance on the mask task for the 3-item, 4-item, and 5-item lists. Performance that is significantly above chance (p < .05) is indicated with an asterisk

		Percent Correct						
Subject	List Length	Sessions	A	B	C	D	E	Total
Benedict	3	3	100*	70*	100*	---	---	70
	4	11	97*	97*	77*	100*	---	72
	5	45	95*	95*	57*	52	100*	27
Horatio	3	4	98*	73*	100*	---	---	72
	4	40	97*	88*	53	100*	---	45
	5	---	---	---	---	---	---	---
Oberon	3	30	100*	72*	100*	---	---	72
	4	40	100*	95*	45.6	100*	---	43
	5	---	---	---	---	---	---	---
Prospero	3	22	98*	71*	100*	---	---	70
	4	40	90*	69*	59	100*	---	37
	5	---	---	---	---	---	---	---

Discussion

Overall, our findings were similar to those reported by Beran et al. (2004). All four subjects were less accurate on BC switch trials, one subject was less accurate on CD switch trials, and there was no impairments in accuracy on DE switch trials. With respect to latencies, all four subjects showed longer response latencies to item B on BC switch trials, two subjects showed longer response latencies to item C on CD switch trials, and one subject showed longer response latencies to item D on DE switch trials. The comparable performance of our subjects and Beran et al.'s (2004) subjects shows that the relatively poor planning abilities displayed by their monkeys cannot be attributed to the use of a joystick. At the very most, our data show that monkeys can plan one or at best two steps ahead. This finding is surprising when you consider the view held by many that monkeys formulate a response plan before executing a simultaneous chain (D'Amato and Colombo, 1988; Kawai, 2001; Ohshiba, 1997; Swartz, Chen, and Terrace, 1991), a view based mainly on the fact that they show a long latency before responding to item A, followed by short and uniform latencies to respond to the remaining items. The idea is that the long latency to item A reflects the process of formulating and loading into memory the sequence of responses that will have to be made, and the short latencies to the remaining items reflect the execution of those planned responses (D'Amato and Colombo, 1988; Kawai, 2001; Ohshiba, 1997; Swartz et al., 1991).

Experiment 2: Eye Movement during Serial-Order Behaviour

At the conclusion of the first study we were faced with two contradictory lines of evidence. On the one hand, the performance of the monkeys on the switch and masked trials suggests an ability to plan one, or at the very best, two steps ahead. On the other hand, the elevated latency to the first item of a series suggests that they can formulate a plan of as many as five items, and then execute that plan with efficiency, as reflected by the short and uniform latencies to the remaining items. In our next experiment (see Scarf and Colombo, 2009) we decided to further explore the source of this elevated latency to the first item and see if it really reflected the formulation of a response plan.

Method

Two juvenile male monkeys, Olympus and Sid Vicious (*Macaca fascicularis*), served as subjects. At the time these data were collected the two monkeys were part of a study examining neural responses in inferior temporal cortex (Gochin et al., 1994) and so they were implanted with recording devices. Details of the recording devices and surgical procedures can be found in Gochin et al. (1994). The monkeys were on a water restriction schedule, but were given free access to water for 3 hours following a recording session. All testing was conducted in a room with a 40W light bulb. The monkey sat in a standard primate chair 15 cm away from a flat black panel, 61 cm long by 28 cm high, housing five inline stimulus projectors (Model 80-0052-1886-A, Industrial Electronic Engineers, Van Nuys, CA), positioned at the four corners and centre of an 11.5 cm square. In front of each projector was a transparent response key measuring 3.2 cm in diameter. The stimuli consisted of a circle (A), a four-lobed pattern (B), a picture of a monkey face (C), a sixteen-lobed pattern (D), and a vertical line (E).

The geometric forms appeared as white on a black background and the monkey face was in colour. At the distance the subject sat from the screen the entire display subtended 42° of visual angle. The monkey sat in a primate chair surrounded by a magnetic field coil (CNC Engineering, Seattle, WA). Details of the magnetic coil and calibration can be found in Scarf and Colombo (2009). Eye movements were sampled at 25 Hz. Cranberry juice (0.4 ml squirt) served as the reward and was delivered to the monkey's mouth by means of a metal spigot. The monkeys were initially trained in a manner similar to that reported by D'Amato and Colombo (1988). Prior to surgery, they were first trained to respond to stimulus A when it was displayed. They were then trained to respond to stimuli A and B in the order A → B. They were then trained on the A → B → C, then A → B → C → D, and finally A → B → C → D → E series. Each stage of training began with a "successive" phase in which only the items of the previous series were displayed at the start of a trial. For example, training on the A → B → C → D → E series began with only stimuli A, B, C, and D displayed, and after correctly responding to these stimuli, stimulus E then appeared. The criterion for the successive phase was one session at 70% correct. They were then trained on the simultaneous phase where all the stimuli were displayed at the same time. The criterion for the simultaneous phase was two sessions at or better than 70% correct. A session consisted of 50 trials.

Results

Postoperatively, Olympus' and Sid Vicious' eye movements were recorded over eight and seven 50-trial sessions, respectively. The eye movement analysis was based on the three highest-performance sessions for each subject. Over these sessions Olympus performed at 80%, 90%, and 94% correct, and Sid Vicious performed at 62%, 66%, and 76% correct. Only data from correct trials was analyzed. We first examined the overall latency pattern. The latency to respond to each of the five items is shown in Figure 2. For Olympus, the long latency to respond to item A was followed by short and uniform latencies to respond to the remaining items B, C, D, and E. The same was generally true for Sid Vicious where the long latency to respond to item A was followed by a shorter latency to respond to item B, and then short and uniform latencies to respond to items C, D, and E. A one-way ANOVA with item (5: A vs B vs C vs D vs E) as a repeated measure was applied to each monkey's data, followed by Student-Neuman-Keuls post-hoc test (evaluated at $p < .05$). For Olympus there was a significant effect of item, $F(4, 524) = 55.2$, $p < .001$, and posthoc tests revealed two non-overlapping subsets, one consisting of the latency to item A, and the other consisting of the latencies to items B, C, D, and E. For Sid Vicious there was also a significant effect of item, $F(4, 400) = 147.7$, $p < .001$, and posthoc tests revealed three non-overlapping subsets, one consisting of the latency to item A, another of the latency to item B, and a third consisting of the latencies to items C, D, and E.

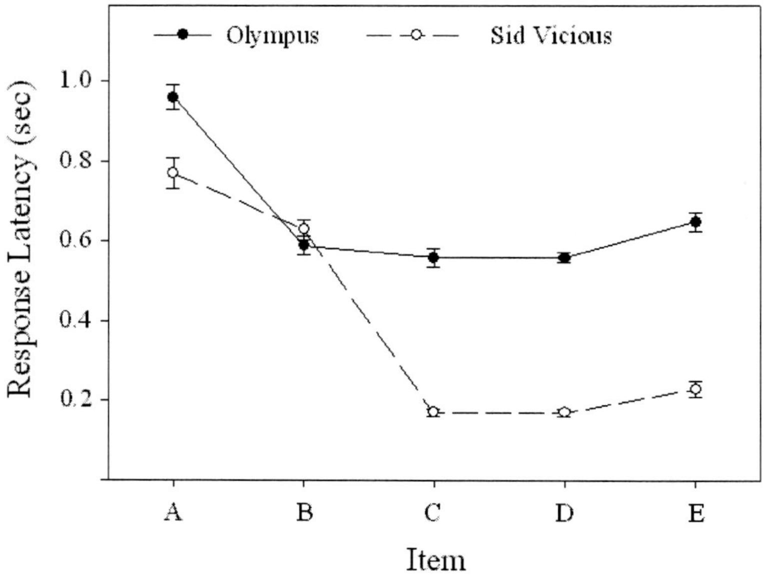

Figure 2. Overall latency to respond to each of the five items for Olympus and Sid Vicious. The overall latency is composed of the time spent scanning items other than the one to which a response is required, the time spent between items but not looking at those items, and the time spent looking at an item before responding to it. The data are based on correct trials only.

The overall latency patterns are either exactly (Olympus) or nearly exactly (Sid Vicious) the patterns used to argue that subjects are planning a sequence of responses (Ohshiba, 1997). According to this view, the elevated latency to item A represents the animal scanning the display and planning the sequence of responses, and the short and uniform latencies to the remaining items represent the execution of the plan. Attractive as this notion is, however, it is not direct evidence for planning. We therefore also undertook a finer analysis of the data to see if the overall latency pattern really is indicative of planning. Several aspects of the scanning data, however, were not consistent with the notion of planning. First, on not a single one of the 132 correct trials analyzed for Olympus, or the 101 correct trials analyzed for Sid Vicious, did the monkey scan all of the list items before responding to stimulus A. Second, if the monkeys were planning we would expect that after locating item A they should search for at least some other items prior to making a response to A (indeed, this is what the elevated latency to item A effectively represents). However, this was rare, and occurred for only 5 of Olympus' 132 correct trials, and 10 of Sid Vicious' 101 correct trials. In other words, once the animal located item A, they responded to it.

Finally, we examined whether seeing an item before responding to it imparted any latency advantage to responding to that item when it was later encountered. In other words, if item B was seen during the search for item A, would the animal "store" the position of item B in memory so that a response to it later (after responding to item A) would be quicker than if they had not seen item B while searching for item A. Figure 3 shows the latency to each of five items for both subjects as a function of whether they had previously seen (s) or not seen (ns) each item before responding to them. A two-way ANOVA with position (4: B vs C vs D vs E) and seen (2: seen vs not seen) as factors was applied to the data. There was a significant effect of position for both Olympus, $F(3, 144) = 3.99$, $p < .05$, and Sid Vicious, $F(3, 78) = 113.52$, $p < .001$, a finding which means nothing more than the fact that the overall latency to the four items differed. More importantly, the latency to respond to an item was not influenced by whether the animal had seen that items before or not, and this was true for both Olympus, $F(1, 48) = .21, p = .65$, and Sid Vicious, $F(1, 26) = 1.05, p = .32$.

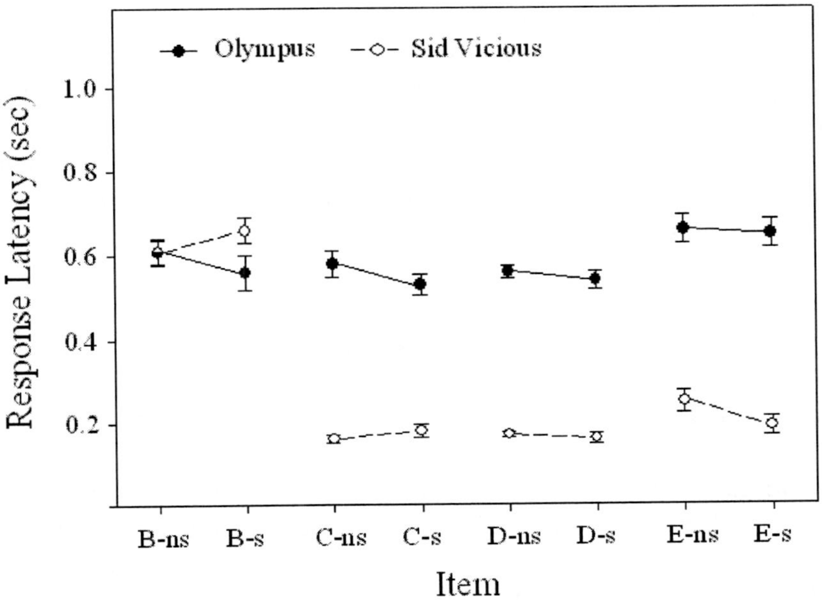

Figure 3. Response latencies to an item as a function of whether that item had not been previously seen (ns) or had been previously seen (s). The data are based on correct trials only.

Discussion

The overall response latencies of our monkeys were exactly (Olympus) or nearly exactly (Sid Vicious) the profile that others have argued is evidence for planning abilities (Kawai, 2001; Ohshiba, 1997; Swartz et al., 1991), that is, a long latency to item A (or A and B for Sid Vicious) followed by relatively short and uniform latencies to the remaining items. Despite this, aspects of the data were clearly inconsistent with the notion of planning. Specifically, there was not a single trial where either monkey scanned all of the list items before responding to item A. Instead, if anything, our data indicated that the monkeys responded to item A once it was located. Additionally, seeing a stimulus prior to responding to it conferred no advantage to responding to that stimulus at later time. We would argue that "loading up" stimuli into memory is a critical component of planning, and we clearly found no evidence of this in the present study.

The obvious question then is: *If the animals were not planning, then what does the elevated response latency to item A represent?* We believe that the elevated response latency to item A is merely an artifact of the behavioural task and the method of stimulus presentation. Simply put, the greater the ambiguity associated with when the stimuli will appear on the screen, the greater the latency to respond to item A. For example, in the

Ohshiba (1997) study, onset of the display was under control of the monkeys, who pressed a black rectangle to make the stimuli appear. Given that the monkeys effectively turned on display themselves, one could argue that there is no ambiguity as to when the stimuli will appear, and indeed the Ohshiba (1997) study reported the shortest A-elevation latency of all the studies, about 330 msec. In contrast, in the Swartz et al. (1991), D'Amato and Colombo (1988), and the current study, the stimuli appeared after the onset of a tone or light that was not under control of the animals. Given that the onset of the tone or light was not under their control, these studies had more ambiguity as to when the stimuli would appear than the Ohshiba (1997) study, and consequently the A-elevation latencies were also somewhat longer (500 msec - 3 sec). In summary, uncertainty with respect to the onset of the display can account for much of the elevated latency to item A. There is no need to resort to explanations based on planning abilities.

General Discussion

Summary of Findings

The performance of our monkeys in Experiments 1 and 2 indicates that they posses only very rudimentary planning abilities. In Experiment 1, the performance of all four monkeys was impaired when the positions of the second and third items were exchanged (BC switch). When the third and fourth items were exchanged (CD switch), however, only one subject continued to show performance deficits, and when the fourth and fifth items were exchanged (DE switch) no subject showed performance deficits. The performance on the mask task was in line with what we observed on the BC, CD, and DE switch conditions. On the 3-item mask task, all four subjects were able reach the criterion of 70% or more correct responses to each masked item. In contrast, on the 4-item mask only Benedict was successful. When later trained on the 5-item mask task, Benedict was able to accurately respond to the first two masked items (B and C) but fell to chance on the fourth masked item (D). The findings of Experiment 1 suggest that monkeys plan one or at most two responses ahead.

In Experiment 2, both monkeys displayed the response latency pattern others have found and used to argue that monkeys are planning a sequence of responses at the outset of a trial (Ohshiba, 1997). However, an analysis of our subject's eye movements provided no support for the planning hypothesis. The findings of Experiment 2, together with those of Experiment 1, demonstrates that at least with respect to the performance of monkeys on the simultaneous chaining paradigm, very little planning is taking place.

The poor planning abilities displayed by our monkeys and Beran et al.'s (2004) monkeys contrasts with the well developed planning abilities displayed by Matsuzawa and colleagues' chimpanzees (Inoue and Matsuzawa, 2007, 2009; Ohshiba, 1997). Remember that with respect to the mask task, Matsuzawa and colleagues' chimpanzees were able to respond to up to eight masked items, a performance level that is well beyond that shown by any monkey that has been tested to date on the mask task. What accounts for the marked difference in performance between chimpanzees and monkeys on the mask task? One possibility is that chimpanzees have better working memory capacity than monkeys, allowing them to remember the locations of a greater number of masked items. Given that monkeys and chimpanzees perform at comparable levels on other tests of memory such as serial probe recognition (Buchanan, Gill, and Braggio, 1981; Kano, Tanaka, and Tomonaga, 2008; Sands and Wright, 1980) and delayed response tasks (Finch, 1942; Nissen, Riesen, and Nowlis, 1938; Riesen and Nissen, 1942; Simpson and Harlow, 1944), a limited memory capacity hypothesis seems unlikely.

A second possibility is that there is, in fact, no difference in the planning abilities of monkeys and chimpanzees. Why then do the monkeys perform so poorly compared to the chimpanzees? Fragaszy and colleagues have suggested that the differences in the planning abilities of chimpanzees and monkeys may simply be the result of differences in their level of vigilance during tasks (Fragaszy et al., 2009). That is, because monkeys regularly pause mid-task and scan their environment they are liable to getting distracted. Tinklepaugh (1921) put forward a similar idea suggesting the difference in performance between chimpanzees and monkeys on memory tasks may be due to greater distractibility of monkeys (see also Hopkins and Washburn, 2002; Spinozzi, 1996). Ettlinger (1983) referred to these differences between the behaviour of chimpanzees and monkeys as a difference in 'cognitive style.' Ettlinger (1983) noted that "the ape's cognitive style differs qualitatively from that of the monkey: not in the direction of greater capacity (using this term as a measure of continuously increasing ability), but in the way that the ape applies himself" (pp. 18).

Whether "cognitive style" is the source of the monkey's planning problems remains to be seen. Certainly a monkey's "cognitive style" does not interfere with their ability to play the serial-order task, which lies at the heart of most planning studies. And certainly monkeys generate latency functions that are, according to some, consistent with the notion of planning, a view that we seriously challenged in our second experiment. However, we are encouraged by the findings of Raby et al. (2007), who showed that birds have remarkable planning abilities when tested in situations similar to their natural foraging style. There is hope, therefore, that with a proper test, monkeys may indeed show planning abilities on par with chimpanzees. At the present, however, there is little support for the notion that monkeys can plan more than one, or at best two, responses ahead.

REFERENCES

Beran, M. J., Pate, J. L., Washburn, D. A., and Rumbaugh, D. M. (2004). Sequential responding and planning in chimpanzees (*Pan troglodytes*) and rhesus macaques (*Macaca mulatta*). *Journal of Experimental Psychology: Animal Behavior Processes, 30*, 203-212.

Biro, D., and Matsuzawa, T. (1999). Numerical ordering in a chimpanzee (*Pan troglodytes*): Planning, executing, and monitoring. *Journal of Comparative Psychology, 113*, 178-185.

Bitterman, M. E. (1965). The evolution of intelligence. *Scientific American, 212*, 92-100.

Brannon, E. M., Cantlon, J. F., and Terrace, H. S. (2006). The role of reference points in ordinal numerical comparisons by rhesus macaques (*Macaca mulatta*). *Journal of Experimental Psychology: Animal Behavior Processes, 32*, 120-134.

Brannon, E. M., and Terrace, H. S. (2000). Representation of the numerosities 1-9 by rhesus macaques (*Macaca mulatta*). *Journal of Experimental Psychology: Animal Behavior Processes, 26*, 31-49.

Buchanan, J. P., Gill, T. V., and Braggio, J. T. (1981). Serial position and clustering effects in chimpanzee's "free recall." *Memory and Cognition*, *9*, 651-660.

Cook, P. and Wilson, M. (2010). Do young chimpanzees have extraordinary working memory? *Psychonomic Bulletin and Review, 17*, 599-600.

D'Amato, M. R., and Colombo, M. (1988). Representation of serial order in monkeys (*Cebus apella*). *Journal of Experimental Psychology: Animal Behavior Processes, 14*, 131-139.

Ettlinger, G. (1983). A comparative evaluation of the cognitive skills of the chimpanzee and the monkey. *International Journal of Neuroscience, 22*, 7-20.

Finch, G. (1942). Delayed matching-from-sample and non-spatial delayed response in chimpanzees. *Journal of Comparative Psychology, 34*, 315-319.

Fragaszy, D. M., Kennedy, E., Murnane, A., Menzel, C., Brewer, G., et al. (2009). Navigating two-dimensional mazes: chimpanzees (*Pan troglodytes*) and capuchins (*Cebus apella sp.*) profit from experience differently. *Animal Cognition, 12*, 491-504.

Gochin, P. M., Colombo, M., Dorfman, G. A., Gerstein, G. L., and Gross, C. G. (1994). Neural ensemble coding in inferior temporal cortex. *Journal of Neurophysiology, 71*, 2325-2337.

Hopkins, W. D., and Washburn, D. A. (2002). Matching visual stimuli on the basis of global and local features by chimpanzees (*Pan troglodytes*) and rhesus monkeys (*Macaca mulatta*). *Animal Cognition, 5*, 27-31.

Inoue, S., and Matsuzawa, T. (2007). Working memory of numerals in chimpanzees. *Current Biology, 17*, 1004-1005.

Inoue, S., and Matsuzawa, T. (2009). Acquisition and memory of sequence order in young and adult chimpanzee (*Pan troglodytes*). *Animal Cognition, 12*, S59-S69.

Iwai, E., Yaginuma, S., and Mishkin, M. (1986). Acquisition of discrimination learning of patterns identical in configuration in macaques (*Macaca mulatta and M. fuscata*). *Journal of Comparative Psychology, 100*, 30-36.

Kano, F., Tanaka, M., and Tomonaga, M. (2008). Enhanced recognition of emotional stimuli in the chimpanzee (*Pan troglodytes*). *Animal Cognition, 11*, 517-524.

Kawai, N. (2001). Ordering and planning in sequential responding to Arabic numerals by a chimpanzee. *Psychologia, 44*, 60-69.

Kawai, N., and Matsuzawa, T. (2000). Numerical memory span in a chimpanzee. *Nature, 403*, 39-40.

Kornell, N., and Terrace, H. S. (2007). The generation effect in monkeys. *Psychological Science 18*, 682-685.

Mulcahy, N. J. and Call, J. (2006). Apes save tools for future use. *Science, 312*, 1038-1040.

Nissen, H. W., Riesen, A. H., and Nowlis, V. (1938). Delayed response and discrimination learning by chimpanzees. *Journal of Comparative Psychology, 26*, 361-386.

Ohshiba, N. (1997). Memorization of serial items by Japanese monkeys, a chimpanzees, and humans. *Japanese Psychological Research, 39*, 236-252.

Paxton, R., and Hampton, R. R. (2009). Tests of planning and the Bischof-Köhler hypothesis in rhesus monkeys (*Macaca mulatta*). *Behavioural Processes, 80*, 238-246.

Raby, C. R., Alexis, D. M., Dickinson, A., and Clayton, N. S. (2007) Planning for the future by western scrub-jays. *Nature, 445*, 919-921.

Riesen, A. H., and Nissen, H. W. (1942). Non-spatial delayed response by the matching technique. *Journal of Comparative Psychology, 34*, 307-313.

Sands, S. F., and Wright, A. A. (1980). Serial probe recognition performance by a rhesus monkey and a human with 10-item and 20-item lists. *Journal of Experimental Psychology: Animal Behavior Processes, 6*, 386-396.

Scarf, D., and Colombo, M. (2009). Eye movements during list execution reveal no planning in monkeys. *Journal of Experimental Psychology: Animal Behavior Processes*, *35*, 587-592.

Scarf, D., and Colombo, M. (2010). The formation and execution of sequential plans in pigeons (*Columba livia*). *Behavioural Processes, 83*, 179-182.

Scarf, D., Danly, E., Morgan, G., Colombo, M., and Terrace, H. (in press). Sequential planning in rhesus monkeys *(Macaca mulatta*). *Animal Cognition.*

Silberberg, A., and Kearns, D. (2009). Memory for the order of briefly presented numerals in humans as a function of practice. *Animal Cognition, 12*, 405-407.

Simpson, M. M., and Harlow, H. F. (1944). Solution by rhesus monkeys of a non-spatial delayed response to the color or form attribute of a single stimulus (Weigl principle delayed reaction). *Journal of Comparative Psychology, 37*, 211-220.

Spinozzi, G. (1996). Categorization in monkeys and chimpanzees. *Behavioural Brain Research, 74*, 17-24.

Subiaul, F., Cantlon, J. F., Holloway, R. L., and Terrace, H.S. (2004). Cognitive imitation in rhesus macaques. *Science, 305*, 407-410.

Swartz, K. B., Chen, S., and Terrace, H. S. (1991). Serial learning in rhesus monkeys: I. Acquisition and retention of multiple four item lists. *Journal of Experimental Psychology: Animal Behavior Processes, 17*, 396-410.

Terrace, H. S. (1984). Simultaneous chaining: The problem it posses for traditional chaining theory. In M. L. Commons, R. J. Herrnstein, and A. R. Wagner (Eds.), *Quantitative analyses of behaviour: Discrimination processes* (pp. 115-138). Cambridge: Ballinger.

Terrace, H. S., Son, L. K., and Brannon, E. M. (2003). Serial expertise of rhesus macaques. *Psychological Science, 14*, 66-73.

Tinklepaugh, O. L. (1932). The multiple delayed reaction with chimpanzees and monkeys. *Journal of Comparative Psychology, 13*, 207-243.

In: Monkeys: Biology, Behavior and Disorders
Editor: Rachel M. Williams, pp 151-165

ISBN: 978-1-61209-911-8

Chapter 7

NEUROPEPTIDES IN THE MONKEY BRAINSTEM

***Ewing Duque*[1], *Arturo Mangas*[2], *Zaida Díaz-Cabiale*[3], *José Angel Narváez*[3] *and Rafael Coveñas*[2,*]**

[1]Pontificia Bolivariana-Montería University, Basic Sciences Center, Laboratory of Neurosciences (Lab. 143), Montería, Colombia

[2]Institute of Neurosciences of Castilla y León (INCYL), Laboratory of Neuroanatomy of the Peptidergic Systems (Lab. 14), Salamanca, Spain

[3]University of Málaga, School of Medicine, Department of Physiology, Málaga, Spain

ABSTRACT

Monkeys are widely used in the laboratory as an experimental animal model in order to answer several scientific questions related to neuroanatomy, neurophysiology, neuropharmacology, neurology and behavior. Brainstem is the region in which the regulation of reflexes and "unconscious" mechanisms (pain transmission, cardiovascular, respiratory…) is located. These mechanisms are mediated by chemical substances such as neuropeptides. These substances, which show a widespread distribution in both the central and peripheral nervous systems, are neuroactive substances that acting as neurotransmitters, neuromodulators (paracrine and autocrine actions) and neurohormones are involved in numerous physiological actions. In the last thirty years, the knowledge on the distribution and functions of the neuropeptides has increased notably in the mammalian central nervous system. Thus, our aim here is to review currently available morphological and physiological data on neuropeptides in the monkey brainstem. We shall thus discuss the following aspects: 1) The distribution of the neuropeptides in the monkey brainstem; 2) The anatomical relationship among the different neuropeptides in the monkey brainstem; 3) The coexistence of neuropeptides in the monkey brainstem; 4) The peptidergic pathways in the monkey brainstem; 5) The physiological functions of the neuropeptides in the monkey brainstem.

Keywords: Neuropeptides; Mesencephalon; Pons; Medulla oblongata; Monkey.

[*]E-mail: covenas@usal.es, Fax number: 34-923294549, Phone number: 34-923294400 extension 1856

Introduction

The brainstem is divided from caudal to rostral region into the medulla oblongata, the pons and the mesencephalon. The brainstem has been implicated in sleep, vocalization, eye movement, pain, analgesia, heart rate, and in sexual, visual, attentive, auditive, motor, respiratory and cardiovascular mechanisms [21, 26, 35]. The brainstem receives somatic and visceral inputs and its neurons send motor efferences by means of the cranial nerves, which innervate the head, neck and sensory organs [35]. In the brainstem there are many specific nuclei, and it also contains the reticular formation, which is involved in several functions.

The monkey has been used as an experimental animal model in neuroanatomy, neurophysiology, neuropharmacology and behavioral studies. The knowledge on the distribution and physiological functions of neuropeptides in the monkey brain has been notably increased, although, at the present, there are many gaps on the distribution of these neuroactive substances in the central nervous system of non-human primates. Here, our aim is to review, in the brainstem of the monkey, the morphological and physiological data currently available about neuropeptides in this region of the monkey central nervous system.

Distribution of Neuropeptides in the Monkey Brainstem

The distribution of some neuropeptides in the monkey brainstem is shown in Tables 1 and 2. There are three families of opioid peptides according to the precursors of such peptides: (1) α-, γ-endorphin, β-endorphin (1-31), β-endorphin (1-27), as well as the non-opiate peptide adrenocorticotropin hormone (ACTH) are produced from pro-opiomelanocortin; (2) methionine-enkephalin, methionine-enkephalin-Arg^6-Phe^7, methionine-enkephalin-Arg^6-Phe^7-Leu^8, leucine-enkephalin, bovine adrenal medullary opioid dodecapeptide, peptide F, synenkephalin from pro-enkephalin; and (3) dynorphin A (1-17), dynorphin A (1-8), leumorphin, dynorphin B (rimorphin), and α-, β-neo-endorphin from pro-dynorphin [see 10]. By using immunohistochemical and radioimmunoassay techniques several opiate peptides have been mapped or located in the monkey brainstem [10, 12, 14, 16-20, 22, 24, 27]. Thus, using immunohistochemical techniques, immunoreactive fibers and cell bodies have been described in the brainstem of several monkey species (*Saimiri sciureus, Macaca fascicularis, Macaca fuscata, Macaca mulata*). For example, cell bodies containing enkephalin (leucine-enkephalin or methionine-enkephalin) have been described in the following nuclei: nucleus of the spinal trigeminal tract, nucleus of the solitary tract, mesencephalic reticular formation, parabrachial nucleus, inferior olivary nucleus, lateral reticular nucleus, mesencephalic periaqueductal gray matter, superior central nucleus, dorsal nucleus of the raphe, inferior colliculus, and in the interpeduncular nucleus (Table 1). Fibers containing enkephalin were observed in the substantia nigra, mesencephalic periacueductal gray matter, interpeduncular nucleus, dorsal nucleus of the raphe, locus coeruleus, and in the parabrachial nucleus (Table 1). Fibers containing methionine-enkephalin-Arg^6-Phe^7-Leu^8 or β-endorphin were observed in the mesencephalic periaqueductal gray matter, interpeduncular nucleus, mesencephalic reticular formation, substantia nigra, dorsal nucleus of the vagus, cuneate nucleus, brachium conjunctivum, bulbar reticular formation, nucleus of the solitary tract, pontine reticular formation, locus coeruleus, lateral cuneate nucleus, inferior olivary nucleus, superior central nucleus and in the gracile nucleus (Table 1). Immunoreactive fibers containing alpha-neo-endorphin were located in the nucleus of the solitary tract, mesencephalic and bulbar reticular formations, interpeduncular nucleus, oculomotor nucleus, annular nucleus, superior central nucleus, reticulotegmental nucleus of the pons, inferior and superior colliculi, principal trigeminal nucleus, locus coeruleus, superior vestibular nucleus, medial vestibular nucleus, lateral vestibular nucleus, inferior vestibular nucleus, prepositus nucleus, hypoglossal nucleus, gracile nucleus, and in the cuneatus nucleus (Table 1). Dynorphin A and dynorphin B have been detected in the substantia nigra, by using radioimmunoassay, whereas immunoreactive structures (fibers/cell bodies) containing dynorphin A and/or dynorphin B were observed in the substantia nigra, parabrachial nucleus, nucleus of the spinal trigeminal tract, and in the nucleus of the solitary tract (Table 1). Autoradiographic studies have shown the presence of receptors for opioid peptides in the brainstem (e.g., interpeduncular nucleus, mesencephalic periaqueductal gray matter, parabrachial nucleus, dorsal nucleus of the raphe, locus coeruleus) [23, 24, 36].

Immunoreactive fibers containing adrenocorticotropin hormone have been described in the mesencephalic periaqueductal gray matter, superior central nucleus, substantia nigra, mesencephalic and pontine reticular formations, inferior colliculus, brachium conjunctivum, nucleus of the spinal trigeminal tract, lateral vestibular nucleus, dorsal nucleus of the raphe, parabrachial nucleus, reticulotegmental nucleus of the pons, lateral reticular nucleus, parabrachial nucleus, and in the gracile nucleus (Table 2) [10]. Tachykinin peptides include neurokinin A, neurokinin B and substance P. It is known that these three neuropeptides have a common C-terminal amino acid sequence and that neurokinin A and substance P are derived from the preprotachykinin A gene, whereas neurokinin B is derived from the preprotachykinin B gene. It is also known that the biological actions of neurokinin A, neurokinin B and substance P are mediated by three receptors, named neurokinin-l, neurokinin-2 and neurokinin-3. Tachykinins have been implicated in several physiological actions such as salivation, the regulation of smooth muscle contraction, depolarization of central neurons, hyperactivity, interaction with dopaminergic A-10 neurons mediating behavioral activation, regulation of blood pressure, and the transmission of the baroreceptor reflex [see 25]. In addition, a loss of tachykinin-containing neurons has been described in neurodegenerative diseases, suggesting that the decrease in the amount of tachykinins could be involved in these and other diseases. Autoradiography methods have shown the presence of substance P receptor binding sites in the monkey gracile nucleus, cuneate nucleus, parabrachial nucleus and in the substantia nigra [34] and by using immunocytochemical procedures the presence of substance P has been reported in the monkey in the substantia nigra, cuneate nucleus, gracile nucleus, inferior colliculus, mesencephalic periaqueductal gray matter, parabrachial nucleus and in the superior colliculus (Table 2) [11, 31]. In the Formosan monkey (*Macaca cyclopsis*), it has been reported the presence of fibers containing FMRF-amide in the substantia nigra, superior colliculus, interpeduncular nucleus, oculomotor nucleus, mesencephalic periaqueductal gray matter, hypoglossal nucleus and in the dorsal nucleus of the raphe (Table 2) [7]. In the rhesus monkey, cholecystokinin receptors have been reported in the mesencephalic periaqueductal gray matter, superior olivary nucleus and in the medial vestibular nucleus (Table 2) [28]. Moreover, by radioimmunoassay the distribution of both cholecystokinin and vasoactive intestinal peptide has been reported in the rhesus monkey [4]. In *Macaca fascicularis*, immunoreactive neurons and/or fibers containing somatostatin have been located in the interpeduncular nucleus, mesencephalic periaqueductal gray matter, brachium conjunctivum, superior central nucleus, substantia nigra, trochlear nucleus, bulbar, pontine and mesencephalic reticular formations, nucleus of the solitary tract, nucleus of the spinal trigeminal tract, inferior colliculus, locus coeruleus, medial vestibular nucleus, inferior olivary nucleus, cochlear nucleus, hypoglossal nucleus, lateral vestibular nucleus, superior colliculus, inferior vestibular nucleus and in the gracile nucleus (Table 2) [10]. Somatostatin binding sites have been also reported in the monkey brain [3].

Moreover, it has been reported the presence of neuropeptide Y mRNA in the nucleus of the solitary tract, dorsal motor nucleus of the vagus, nucleus of the spinal trigeminal tract, mesencephalic reticular formation, locus coeruleus, mesencephalic periaquieductal gray matter, dorsal nucleus of the raphe, and in the pontine reticular formation [30]. By using of the immunohistochemical methods, the presence of calcitocin gene-related peptide has been reported in the nucleus of the spinal trigeminal tract, hypoglossal nucleus and in the inferior olive of the monkey (*Macaca fascicularis*) [1, 2, 15]. In this latter species, the presence of immunoreactive structures containing neurokinin A has been reported in the substantia nigra, mesencephalic periaqueductal gray matter, inferior colliculus, pontine reticular formation, nucleus of the solitary tract, interpeduncular nucleus, annular nucleus and in the red nucleus [10]. In *Macaca fascicularis* fibers containing neurotensin have been observed in the mesencephalic periaqueductal gray matter, nucleus of the solitary tract, interpeduncular nucleus, superior central nucleus, substantia nigra, oculomotor nucleus and in the superior and inferior colliculi [10]. Neuropeptide FF receptor 2 has been reported in the nucleus of the spinal trigeminal tract, cuneate nucleus and in the gracile nucleus of the African green monkey [37]; by immunohistochemstry, the presence of orexin in the locus coeruleus and in the brainstem visuomotor areas of the Rhesus monkey has been described [13, 33], whereas the distribution of angiotensin-converting enzime and nociceptin/orphanin FQ binding sites has been reported in the monkey brainstem [5, 6].

Table 1. Distribution of neuropeptides in the monkey brainstem

NP	MET-ENK		MET-8		LEU-ENK			β-END	α-MSH		α-NEO		DYN-A		DYN-B	
Nuclei	F	CB	F	CB	F	CB	F	CB	F	CB	F	CB	F	CB	F	CB
III	-	-	-	-	-	-	-	-	-	-	+	-	-	-	-	-
IV	-	-	-	-	-	-	+	-	-	-	-	-	-	-	-	-
VI	-	-	-	-	-	-	-	-	-	-	-	-	-	-	-	-
VIII	-	-	+	-	-	-	-	-	-	-	-	-	-	-	-	-
X	-	-	+	+	-	-	-	-	-	-	-	-	-	-	-	-
XII	-	-	-	-	-	-	-	-	-	-	+	-	-	-	-	-
Ann	-	-	-	-	-	-	+	-	-	-	+	-	-	-	-	-
BC	-	-	+	-	-	-	+	-	-	-	-	-	-	-	-	-
BP	-	-	-	-	-	-	-	-	-	-	-	-	-	-	-	-
BRTF	-	-	+	-	-	-	+	-	-	-	+	-	-	-	-	-
Cn	-	-	+	-	-	-	+	-	-	-	+	-	-	-	-	-
DBC	-	-	-	-	-	-	-	-	-	-	-	-	-	-	-	-
DR	-	+	-	+	+	-	-	-	-	-	-	-	-	-	-	-

NP / Nuclei	MET-ENK		MET-8		LEU-ENK		β-END		α-MSH		α-NEO		DYN-A		DYN-B	
	F	CB	F	CB	F	CB	F	CB	F	CB	F	CB	F	CB	F	CB
Gr	-	-	+	-	-	-	+	-	-	-	+	-	-	-	-	-
IC	-	+	+	-	-	-	-	-	-	-	+	-	-	-	-	-
IO	-	-	+	-	-	+	+	-	-	-	+	-	-	-	-	-
IP	-	+	+	+	+	-	-	-	-	-	+	-	-	-	-	-
LC	+	+	+	+	+	-	-	-	-	-	+	-	-	-	-	-
LCn	-	-	+	-	-	-	-	-	-	-	-	-	-	-	-	-
LRet	-	-	-	-	-	+	-	-	-	-	-	-	-	-	-	-
ML	-	-	-	-	-	-	+	-	-	-	-	-	-	-	-	-
MRTF	-	+	+	+	-	-	+	-	-	-	+	-	-	-	-	-
NPV	-	-	+	-	-	-	-	-	-	-	+	-	-	-	-	-
NSpV	-	+	-	+	-	-	+	-	-	-	-	+	+	+	+	+
NTS	-	+	+	+	-	-	+	-	-	-	+	-	-	+	-	+
PAG	-	-	+	+	+	-	+	-	+	-	+	-	-	-	-	-
Pb	-	+	-	+	+	-	-	-	-	-	-	-	+	+	+	+
Pp	-	-	+	-	-	-	-	-	-	-	+	-	-	-	-	-
PRTF	-	-	+	-	-	-	+	-	-	-	-	-	-	-	-	-
Py	-	-	+	-	-	-	-	-	+	-	-	-	-	-	-	-
PyD	-	-	-	-	-	-	-	-	-	-	-	-	-	-	-	-
RN	-	-	+	-	-	-	-	-	-	-	-	-	-	-	-	-
RTP	-	-	-	-	-	-	+	-	-	-	+	-	-	-	-	-
SC	-	-	-	-	-	-	-	-	-	-	+	-	-	-	-	-
SCN	-	+	+	-	-	-	+	-	-	-	+	-	-	-	-	-
SN	+	-	+	+	+	-	-	-	-	-	-	-	+	-	+	-
SO	-	-	-	-	-	-	-	-	-	-	-	-	-	-	-	-
TB	-	+	-	+	-	-	-	-	-	-	-	-	-	-	-	-
VesI	-	-	+	-	-	-	-	-	-	-	+	-	-	-	-	-
VesL	-	-	+	-	-	-	-	-	-	-	+	-	-	-	-	-
VesM	-	-	+	-	-	-	+	-	-	-	+	-	-	-	-	-
VesS	-	-	+	-	-	-	-	-	-	-	+	-	-	-	-	-

F: fibers; CB: cell bodies; NP: neuropeptide; +: presence; -: absence or not studied. For the nomenclature of neuropeptides and the brainstem nuclei, see list of abbreviations.

Table 2. Distribution of neuropeptides in the monkey brainstem

NP	FMRF		SP		CCK		SOM		ACTH	
Nuclei	F	CB	F	CB	F	CB	F	CB	F	CB
III	+	-	-	-	-	-	-	-	-	-
IV	-	-	-	-	-	-	+	-	-	-
VI	-	-	-	-	-	-	-	-	-	-
VIII	-	-	-	-	-	-	+	-	-	-
X	-	-	-	-	-	-	-	-	-	-
XII	+	-	-	-	-	-	+	-	-	-
Ann	-	-	-	-	-	-	-	-	-	-
BC	-	-	-	-	-	-	+	-	+	-
BP	-	-	-	-	-	-	-	-	-	-
BRTF	-	-	-	-	-	-	+	-	-	-
Cn	-	-	+	-	-	-	-	-	-	-
DBC	-	-	-	-	-	-	-	-	-	-
DR	+	-	-	-	-	-	-	-	+	-
Gr	-	-	+	-	-	-	+	-	+	-
IC	-	-	+	-	-	-	+	-	+	-
IO	-	-	-	-	-	-	+	-	-	-
IP	+	-	-	-	-	-	+	+	-	-
LC	-	-	-	-	-	-	+	-	-	-
LCn	-	-	-	-	-	-	-	-	-	-
LRet	-	-	-	-	-	-	-	-	+	-
ML	-	-	-	-	-	-	-	-	-	-
MRTF	-	-	-	-	-	-	+	-	+	-
NPV	-	-	-	-	-	-	+	-	-	-
NSpV	+	-	-	-	-	-	+	-	+	-
NTS	+	-	-	-	-	-	+	-	+	-
PAG	+	-	+	-	+	-	+	+	+	-
Pb	-	-	+	-	-	-	-	-	+	-
Pp	-	-	-	-	-	-	+	-	+	-
PRTF	-	-	-	-	-	-	+	-	+	-
Py	-	-	-	-	-	-	+	-	-	-
PyD	-	-	-	-	-	-	-	-	-	-
RN	-	-	-	-	-	-	-	-	-	-
RTP	-	-	-	-	-	-	-	-	+	-
SC	+	-	+	-	-	-	+	-	+	-
SCN	-	-	-	-	-	-	+	-	+	-
SN	+	-	+	-	+	-	+	-	+	-
SO	-	-	-	-	+	-	-	-	-	-
TB	-	-	-	-	-	-	-	-	-	-
VesI	-	-	-	-	-	-	+	-	+	-
VesL	-	-	-	-	-	-	+	-	+	-
VesM	-	-	-	-	+	-	+	-	-	-
VesS	-	-	-	-	-	-	-	-	-	-

F: fibers; CB: cell bodies; NP: neuropeptide; +: presence; -: absence or not studied. For the nomenclature of neuropeptides and the brainstem nuclei, see list of abbreviations.

Figure 1. shows the percentages of the monkey brainstem nuclei in which fibers and cell bodies containing different neuropeptides were observed. Thus, 100% correspond to the total number of brainstem nuclei/regions/tracts (n = 42) listed in Figures 1 and 2.

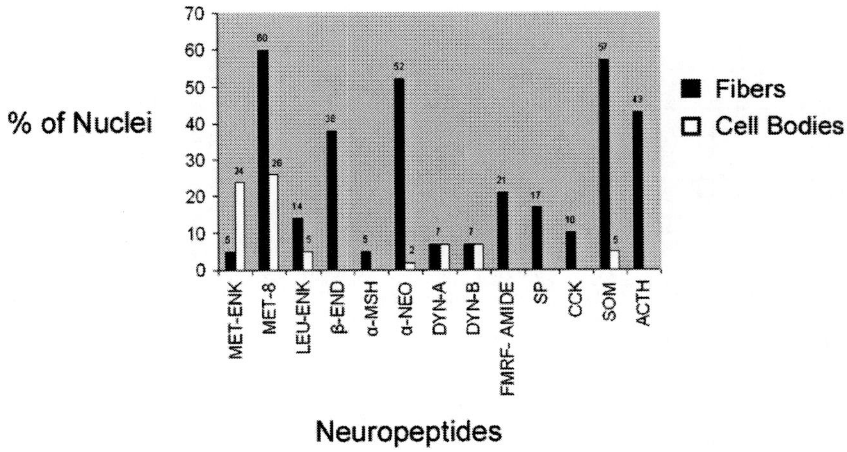

Figure 1. Percentages of brainstem nuclei of the monkey that contain a neuropeptide located in fibers or cell bodies. For the nomenclature of neuropeptides, see list of abbreviations.

Note that the higher percentages of nuclei with immunoreactive fibers correspond to methionine-enkephalin-8 (Met-8) (60%), somatostatin (SOM) (57%), alpha-neo-endorphin (α-NEO) (52%), adrenocorticotropin hormone (ACTH) (43%) and beta-endorphin (β-END) (38%). In the same manner, methionine-enkephalin-8 and methionine-enkephalin (Met-Enk) showed the higher percentages of nuclei containing immunoreactive cell bodies (26% and 24%, respectively).

Anatomical Relationships among the Neuropeptides in the Monkey Brainstem

Table 3. shows the anatomical relationships among the fibers containing neuropeptides located in the monkey brainstem. The percentages showed were calculated taking the total number of the monkey brainstem nuclei (n = 25) in which methionine-enkephalin-8-immunoreactive fibers were observed as 100% and so on for the other neuropeptides (e.g., n = 24 for somatostatin, n = 22 for alpha-neo-endorphin).

Thus, for example, metionine-enkephalin-8 and somatostatin showed a value of 80%. This indicates that in 80% of the monkey brainstem nuclei in which methionine-enkephalin-8-immunoreactive fibers were observed, somatostatin-immunoreactive fibers were also found. The percentage in the case of alpha-neo-endorphin and substance P is 23%.

This value indicates that in 23% of the nuclei in which immunoreactive fibers containing alpha-neo-endorphin were observed, immunoreactive fibers containing substance P were also found.

An anatomical relationship between dynorphin A and alpha-melanocyte-stimulating hormone or between alpha-neo-endorphin and dynorphin B was not observed (percentages: 0%).

Table 3. Anatomical relationship between the fibers containing neuropeptides in the monkey brainstem

	Met8 N=25												
Som	80%	Som N=24											
Neo	68%	71%	Neo N=22										
ACTH	48%	58%	50%	ACTH N=18									
End	44%	50%	50%	50%	End N=16								
FMRF	16%	29%	27%	33%	19%	FMRF N=9							
SP	20%	21%	23%	33%	19%	33%	SP N=7						
Leu-enk	16%	17%	14%	22%	6%	44%	43%	Leu-enk N=6					
CCK	12%	12%	9%	11%	12%	22%	29%	33%	CCK N=4				
DynA	8%	8%	0%	17%	6%	22%	29%	33%	25%	DynA N=3			
DynB	8%	8%	0%	17%	6%	22%	29%	33%	25%	100%	DynB N=3		
MSH	8%	8%	5%	6%	6%	11%	14%	17%	25%	0%	0%	MSH N=2	
Met-enk	8%	8%	5%	6%	0%	11%	14%	33%	25%	33%	33%	0%	Met-enk N=2

For the nomenclature of neuropeptides, see list of abbreviations.

Table 4. Anatomical relationship between the cell bodies containing neuropeptides in the monkey brainstem

	Met8 N=11						
Met-enk	73%	Met-enk N=10					
DynA	27%	30%	DynA N=3				
DynB	27%	30%	100%	DynB N=3			
Som	18%	10%	0%	0%	Som N=2		
Leu-enk	0%	0%	0%	0%	0%	Leu-enk N=2	
Neo	9%	10%	33%	33%	0%	0%	Neo N=1

For the nomenclature of neuropeptides, see list of abbreviations.

Table 4. shows the anatomical relationships among the cell bodies containing neuropeptides located in the monkey brainstem. The percentages showed were calculated taking the total number of the monkey brainstem nuclei (n = 11) in which methionine-enkephalin-8-immunoreactive cell bodies were observed as 100% and so on for the other neuropeptides (e.g., n = 10 for methionine-enkephalin, n = 3 for dynorphin A). The percentage in the case of methionine-enkephalin and somatostatin is 10%. This value indicates that in 10% of the nuclei in which immunoreactive cell bodies containing methionine-enkephalin were observed, immunoreactive cell bodies containing somatostatin were also found.

COEXISTENCE OF NEUROPEPTIDES IN THE MONKEY BRAINSTEM

The presence of several neuropeptides in the same monkey brainstem nuclei indicates a possible interaction among such neuropeptides and an elaborate modulation of functions in which those brainstem nuclei are involved (see Tables 1 and 2). In addition, the localization of several different neuropeptides in the same brainstem nuclei indicates the possibility that two or more of them may coexist in the same neuron. For example, in the monkey, the coexistence of substance P and dynorphin has been found in the substantia nigra [31]. Moreover, the coexistence of neuropeptides (e.g., calcitocin gene-related peptide) with classical neurotransmitters (e.g., serotonin) has been also reported in the monkey brainstem (dorsal nucleus of the raphe, hypoglossal nucleus) [1].

According to the distribution and to the morphological characteristics of the peptidergic cell bodies found in the monkey brainstem (see Tables 1 and 2), the following possible coexistence of neuropeptides should be suggested:

1. Methionine-enkephalin and alpha-neo-endorphin in the nucleus of the spinal trigeminal tract [10, 19].
2. Methionine-enkephalin and somatostatin in the interpeduncular nucleus [10].
3. Methionine-enkephalin-Arg^6-Phe^7-Leu^8 and somatostatin in the interpeduncular nucleus [10].

In order to demonstrate the coexistence of the neuropeptides indicated above, future works should be carried out using double immunofluorescence procedures.

PEPTIDERGIC PATHWAYS IN THE MONKEY BRAINSTEM

Tables 5 and 6 show respectively nuclei of the monkey brainstem in which a moderate/high density of peptidergic fibers, but no peptidergic cell bodies have been observed and nuclei in which a moderate density of immunoreactive cell bodies has been found, but no immunoreactive fibers. These data indicate that the brainstem nuclei shown in Table 5 could receive peptidergic afferents arising from neurons located inside and/or outside the brainstem, whereas the nuclei included in Table 6 could contain projecting neurons, which could send projections to other brainstem nuclei and/or to other parts of the central nervous system. In the future, the possible brainstem peptidergic afferents and brainstem peptidergic projecting neurons, showed in Tables 5 and 6, should be demonstrated.

Table 5. Possible peptidergic afferents to the monkey brainstem nuclei

NP / Nuclei	α-NEO	SOM	MET-8	ACTH	β-END	LEU	FMRF	DYN-A	MET
III	++						++		
IV					++				
XII	++	++							
Ann	++				++				
Cn	++		++						
DR				++		++	++		
Gr		++							
IC		++	++	++					
IP	++					++	++		
LC	++	++							
MRTF		++							
NPV	++	++	++						
NSpV		++			++				
NTS	++	+++		++					
PAG	++			++		++	++		
Pp		++							
PRTF			++						
RN			++						
RTP	++			++	++				
SC	++								
SCN	++		+++	++	++				
SN		++		++		++	++		++
VesL		++							
VesM	++	++							
VesS	++								

For the nomenclature of neuropeptides, see list of abbreviations. ++/+++: moderate/high density of immunoreactive fibers.

Table 6. Possible peptidergic projecting neurons in the monkey brainstem nuclei

NP / Nuclei	MET-8	MET-ENK	DYN-A	DYN-B
DR	++	++		
IC		++		
IP	++	++		
NP / Nuclei	MET-8	MET-ENK	DYN-A	DYN-B
MRTF		++		
NSpV	++	++		
NTS			++	++
Pb	++	++	++	
SCN		++		
TB	++	++		

For the nomenclature of neuropeptides, see list of abbreviations. ++: moderate density of immunoreactive cell bodies.

Physiological Functions of Neuropeptides in the Monkey Brainstem

The widespread distribution of neuropeptides in the monkey brainstem suggests that these substances, acting as neurotransmitters and/or neuromodulators, should be involved in several physiological actions. For instance, in the monkey the presence of immunoreactive structures containing adrenocorticotropin hormone, methionine-enkephalin, alpha-neo-endorphin and methionine-enkephalin-Arg^6-Gly^7-Leu^8 in the nucleus of the solitary tract suggests that these neuropeptides could be involved in respiratory, cardiovascular and gastric regulation, and in the modulation of taste responses [8, 9, 10, 29]. It has been suggested that respiratory disturbances would be associated with a decreased expression of the methionine-enkephalin-Arg^6-Gly^7-Leu^8 in the parabrachial nucleus, dorsal nucleus of the vagus and in the nucleus of the solitary tract [32].

Furthemore, the presence of adrenocorticotropin hormone, dynorphin A, dinorphin B, substance P, cholecytoskinin, and somatostatin in the substantia nigra suggests a possible involment of these neuropeptides in motor function. The presence of leucine-enkephalin and methionine-enkephalin-Arg^6-Gly^7-Leu^8 in the mesencephalic periaqueductal gray matter may be involved in the pain control system [19], whereas the presence in the inferior colliculus of methionine-enkephalin, methionine-enkephalin-Arg^6-Gly^7-Leu^8, adrenocorticotropin hormone, alpha-neo-endorphin and substance P indicates that they are involved in auditory mechanisms.

Future Research on Neuropeptides in the Monkey Brainstem

In the monkey brainstem there is much to be done in order to gain full insight into the distribution and the physiological functions of the neuropeptides in this region of the central nervous system. Thus, in addition to study the distribution of other neuropeptides that, at present, have not been studied in detail in the monkey brainstem, in the future in this central nervous system region *in situ* hybridization techniques should be carried out in order to compare the results obtained using this technique with those described by using immunocytochemistry for the distribution of neuropeptides. In addition, the following aspects should be explored in greater depth in the monkey brainstem since they have not received sufficient attention: 1) The distribution of the receptors for neuropeptides; 2) The distribution of peptidases; 3) The location of cell bodies that originate peptidergic afferences to the brainstem; 4) The projections of the peptidergic neurons observed in the brainstem; 5) Synaptic connections containing neuropeptides; 6) The coexistence of neuropeptides, and 7) The physiological functions of neuropeptides. In sum, the application of other methodologies (combining immunocytochemical and tract-tracing methods, immunocytochemical and electron microscopic methods, immunofluorescence methods…) is required in order to gain further insight into the distribution and functions of the neuropeptides in the monkey brainstem.

ACKNOWLEDGMENTS

This work has been supported by the Spanish DGCYT BFU2005-02241, BFU2008-03369 and Junta de Andalucia SEJ01323.

ABBREVIATIONS

III: oculomotor nucleus
IV: trochlear nucleus
VI: abducens nucleus
VIII: cochlear nucleus
X: dorsal nucleus of the vagus
XII: hypoglossal nucleus
ACTH: adrenocorticotropin hormone
Ann: annular nucleus
BC: brachium conjunctivum
BP: brachium pontis
BRTF: bulbar reticular formation
CCK: cholecystokinin
Cn: cuneate nucleus
DBC: decussation of the brachium conjunctivum
DR: dorsal nucleus of the raphe
DYN A: dynorphin A
DYN B: dynorphin B
β-END: beta-endorphin
FMRF: FMRF-amide
Gr: gracile nucleus
IC: inferior colliculus
IO: inferior olivary nucleus
IP: interpeduncular nucleus
LC: locus coeruleus
LCn: lateral cuneate nucleus
LEU-ENK: leucine-enkephalin
LRet: lateral reticular nucleus
MET-8: methionine-enkephalin-Arg^6-Phe^7-Leu^8
MET-ENK: methionine-enkephalin
ML: medial lemniscus
MRTF: mesencephalic reticular formation
α-MSH: alpha-melanocyte-stimulating hormone
α-NEO: alpha-neo-endorphin
NPV: principal trigeminal nucleus
NPY: neuropeptide Y
NSpV: nucleus of the spinal trigeminal tract
NTS: nucleus of the solitary tract

PAG: mesencephalic periaqueductal gray matter
Pb: parabrachial nucleus
Pp: prepositus nucleus
PRTF: pontine reticular formation
Py: pyramidal tract
PyD: pyramidal decussation
RN: red nucleus
RTP: reticulotegmental nucleus of the pons
SC: superior colliculus
SCN: superior central nucleus
SN: substantia nigra
SO: superior olivary nucleus
SOM: somatostatin
SP: substance P
TB: nucleus of the trapezoid body
VesI: inferior vestibular nucleus
VesL: lateral vestibular nucleus
VesM: medial vestibular nucleus
VesS: superior vestibular nucleus

REFERENCES

[1] Arvidsson, U; Schalling, M; Cullheim, S; Ulfhake, B; Terenius, L; Verhofstad, A; Hökfelt, T. Evidence of coexistence between calcitonin gene-related peptide and serotonin in the bulbospinal pathway in the monkey. *Brain Res.* 1990; 532, 47-57.

[2] Arvidsson, U; Ulfhake, B; Cullheim, S; Terenius, L; Hökfelt, T. Calcitonin gene-related peptide in monkey spinal cord and medulla oblongata. *Brain Res.* 1991; 558, 330-334.

[3] Beal, MF; Tran, VT; Mazurek, MF; Chattha, G; Martin, JB. Somatostatin binding sites in human and monkey brain: localization and characterization. *J. Neurochem.* 1986; 46, 359-365.

[4] Beinfeld, MC; Lewis, ME; Eiden, LE; Nilaver, G; Pert, CB. The distribution of cholecystokinin and vasoactive intestinal peptide in rhesus monkey brain determined by radioimmunoassay. *Neuropeptides* 1983; 3, 337-344.

[5] Bridge, KE; Wainwrifht, A; Reilly, K; Oliver, KR. Autoradiographic localization of ^{125}I[Try^{14}] nociceptin/orphain FQ binding sites in macaque primate CNS. *Neuroscience* 2003; 118, 513-523.

[6] Chai, SY; McKinley, MJ; Paxinos, G; Mendelsohn, FA. Angiotensin converting enzyme in the monkey (*Macaca fascicularis*) brain visualized by *in vitro* autoradiography. *Neuroscience* 1991; 42, 483-495.

[7] Chen, S; Tsai, MS; Shen, CL. Distribution of FMRFamide-like immunoreactivity in the central nervous system of the Formosan monkey (*Macaca cyclopsis*). *Peptides* 1989; 10, 825-834.

[8] Cheng, PY; Liu-Chen, LY; Chen, C; Pickel, VM. Immunolabeling of Mu opioid receptors in the rat nucleus of the solitary tract: extrasynaptic plasmalemmal

localization and association with Leu5-enkephalin. *J. Comp. Neurol.* 1996; 371, 522-536.

[9] Cheng-Shu, L; Davis, BJ; Smith, DV. Opioid modulation of taste responses in the nucleus of the solitary tract. *Brain Res.* 2003; 965, 21-34.

[10] Coveñas R; Duque, E; Mangas, A; Marcos, P; Narváez, JA. Neuropeptides in the monkey (*Macaca fascicularis*) brainstem. In: Mangas A, Coveñas R, Geffard M, editors. *Brain Molecules: from Vitamins to Molecules for Axonal Guidance.* Trivandrum: Transworld Research Network; 2008; 131-156.

[11] DiFiglia, M; Aronin, N; Leeman, SE. Immunoreactive substance P in the substantia nigra of the monkey: light and electron microscopic localization. *Brain Res.* 1981; 233, 2381-2388.

[12] Dores, R; Akil, H. Steady state levels of pro-dynorphin-related end products in the striatum and substantia nigra of the adult rhesus monkey. *Peptides* 1985; 6, 143-148.

[13] Downs, JL; Dunn, MR; Borok, E; Shanabrough, M; Horvath, TL; Kohama, SG; Urbanski, HF. Orexin neuronal changes in the locus coeruleus of the aging rhesus macaque. *Neurobiol. Aging,* 2007; 28, 1286-1295.

[14] Edwards, DL; Poletti, CE; Foote, WE. Evidence for leucine-enkephalin-immunoreactive neurons in the medulla which project to spinal cord in squirrel monkey. *Brain Res.* 1987; 437, 197-203.

[15] Fried, K; Risling, M; Arvidsson, U; Paulie, S. Nerve growth factor receptor-like immunoreactivity in nerve fibers in the spinal and medullary dorsal horn of the adult monkey and cat: correlation with calcitonin gene-related peptide-like immunoreactivity. *Brain Res.* 1990; 536, 321-326.

[16] Haber, S; Elde, R. The distributions of enkephalin-immunoreactive neuronal cell bodies in the monkey brain: preliminary observations. *Neuroscie. Lett.* 1982; 32, 247-252.

[17] Ibuki, T; Okamura, H; Miyazaki, M; Yanaihara, N; Zimmerman, EA; Ibata, Y. Comparative distribution of three opioid systems in the lower brainstem of the monkey (*Macaca fuscata*). *J. Comp. Neurol.* 1989; 279, 445-456.

[18] Inagaki, S; Parent, A. Distribution of substance P and enkephalin-like immunoreactivity in the substantia nigra of rat, cat, and monkey. *Brain Res. Bull.* 1984; 13, 319-329.

[19] Inagaki, S; Parent, A. Distribution of enkephalin-immunoreactive neurons in the forebrain and upper brainstem of squirrel monkey. *Brain Res.* 1985; 359, 267-280.

[20] Inagaki, S; Kubota, Y; Kito, S. Ultrastructural localization of enkephalin-immunoreactivity in the substantia nigra of the monkey. *Brain Res.* 1986; 362, 171-174.

[21] Kandel, R; Schwartz, JH; Jessell, TM. *Principles of Neural Science*. New York: McGraw-Hill; 2000.

[22] Khachaturian, H; Lewis, ME; Haber, SZ; Houghten, RA; Akil, H; Watson, SJ. Prodynorphin peptide immunocytochemistry in rhesus monkey brain. *Peptides* 1985; 6, 155-166.

[23] Lewis, ME; Khachaturian, H; Akil, H; Watson, SJ. Anatomical relationship between opioid peptides and receptors in rhesus monkey brain. *Brain Res. Bull.* 1984; 13, 801-812.

[24] Lewis, ME; Khachaturian, H; Watson, SJ. Comparative distribution of opiate receptors and three opioid peptide neuronal systems in rhesus monkey central nervous system. *Life Sci.* 1983; 33, 239-242.

[25] Marcos, P; Coveñas, R; de León, M; Narváez, JA; Tramu, G; Aguirre, JA; González-Barón, S. Neurokinin A-like immunoreactivity in the cat brainstem. *Neuropeptides* 1993; 25, 105-114.

[26] Martin, GF; Holstege, G; Mehler WR. *Reticular Formation of the Pons and Medulla*. San Diego: Academic Press; 1990.

[27] Matsukura, S; Yoshimi, H; Sueoka, S; Katoka, K; Ono, T; Ohgushi, N. The regional distribution of immunoreactive β-endorphin in the monkey brain. *Brain Res.* 1978; 159, 228-233.

[28] Mercer, LD; Beart, PM; Horne, MK; Finkelstein, DI; Carrive, P; Paxinos, G. On the distribution of cholecystokinin B receptors in monkey brain. *Brain Res.* 1996; 738, 313-318.

[29] Palkovits, M; Eskay, RL. Distribution and possible origin of beta-endorphin and ACTH in discrete brainstem nuclei of rats. *Neuropeptides* 1987; 9, 123-137.

[30] Pau, K; Yu, J; Lee CJ; Spies, HG. Topographic localization of neuropeptide Y mRNA in the monkey brainstem. *Regul. Peptides.* 1998; 75-76, 145-153.

[31] Reiner, A; Medina, L; Habert, N. The distribution of dynorphinergic terminals in striatal target regions in comparison to the distribution of substance P-containing and enkephalinergic terminals in monkeys and humans. *Neuroscience* 1999; 88, 775-793.

[32] Saito, Y; Ito, M; Ozawa, Y; Matsuishi, T; Hamano, K; Takashima, S. Reduced expression of neuropeptides can be related to respiratory disturbances in Rett syndrome. *Brain Develop.* 2001; 1, 122-126.

[33] Schreyer, S; Büttner-Ennever, JA; Tang, X; Mustari, MJ; Horn, AK. Orexin-A inputs onto visuomotor cell groups in the monkey brainstem. *Neuroscience* 2009; 164, 629-640.

[34] Schwark, HD; Petit, MJ; Fuchs, JL. Distribution of substance P receptor binding in dorsal column nuclei of rat, cat, monkey, and human. *Brain Res.* 1998; 786, 259-262.

[35] Shepherd G. *Neurobiology*. New York: Oxford University Press; 1988.

[36] Sim-Selley, LJ; Daunais, JB; Porrino, LJ; Childersi, SR. Mu an Kappa opioid-stimulated [^{35}S] guanylyl-5′-o-(γ-THIO)-triphosphate binding in cynomolcus monkey brain. *Neuroscience* 1999; 94, 651-662.

[37] Zeng, Z; McDonald, TP; Wang, R; Liu, Q; Austin, CP. Neuropeptide FF receptor 2 (NPFF2) is localized to pain-processing regions in the primate spinal cord and the lower level of the medulla oblongata. *J. Chem. Neuroanat.* 2003; 25, 269-278.

In: Monkeys: Biology, Behavior and Disorders
Editor: Rachel M. Williams, pp 167-179

ISBN: 978-1-61209-911-8

Chapter 8

Developmental Neuronal Toxicity and the Rhesus Monkey

Cheng Wang*, Merle G. Paule, Fang Liu, Xuan Zhang, Tucker A. Patterson and William Slikker

Division of Neurotoxicology, National Center for Toxicological Research, U.S. Food and Drug Administration, 3900 NCTR Rd., Jefferson, AR 72079, USA

The rhesus monkey, *Macaca mulatta,* is an animal model used to inform aspects of human physiology, pathology, pharmacology, toxicology, and systems biology. Because of obvious limitations it is not possible to thoroughly explore the effects of pediatric anesthetic agents on neurons in human infants or children, nor is it possible to determine the dose-response or time-course for potential anesthetic-induced neuronal cell death in humans.

Due to the complexity of the primate brain, the monkey is often the animal model of choice for neurotoxicological experiments and it is in the rhesus monkey that the phenomenon of interest (anesthetic-induced neuronal cell death in the brain) has been previously reported.

In addition, the anatomical and functional characterization of the monkey central nervous system (CNS) is extensive, thereby facilitating the interpretation of findings using this model. The relevance of the anesthetic-induced neuronal cell death observed in rodent models to children is inferred if similar effects can be observed in a developing nonhuman primate. Thus, the present review will focus on research findings from studies using the developing rhesus monkey to address: (1) the evidence for, and characteristics of, anesthetic-induced neuronal cell death in the developing monkey brain; (2) pharmacodynamic outcomes and physiological parameters associated with anesthetic-induced developmental neurotoxicity; (3) relationships between anesthetic-induced neurotoxicity and developmental stage at time of exposure; (4) how the complex behavioral capabilities of the rhesus monkey provide opportunities to study the effects of anesthetic exposure during development on acquisition of important behaviors such as learning and memory that have direct relevance to humans; and

* Tel.: 870-543-7259; fax: 870-543-7745. E-mail address: cheng.wang@fda.hhs.gov

(5) the use of a nonhuman primate model to decrease the uncertainty in extrapolating pre-clinical data to the human condition.

Keywords: nonhuman primate; NMDA receptor antagonist; ketamine; developmental neurotoxicity; brain function.

INTRODUCTION

The rhesus monkey, *Macaca mulatta,* is commonly used to study aspects of biology it shares with humans. Due to the complexity of the primate brain, the monkey is often the model of choice for neurological and behavioral experiments [1]. Recent investigations in rodents have shown that some anesthetic drugs induce abnormal levels of neurodegeneration (apoptosis) if administered during critical periods of brain development [2-4]. Because it is not possible to thoroughly explore the effects of pediatric anesthetic agents on neuronal cell death or to determine the dose-response or time-course of potential anesthetic-induced neuronal cell death in human infants or children, the nonhuman primate offers as close an animal model as possible. The anatomical and functional complexity of the monkey CNS facilitates the interpretation of data with respect to the extrapolation of findings to humans. In addition, no other commonly used research animal has a functional fetal-placental unit, a propensity for single births and a fetal-to-maternal weight ratio comparable to that of humans. Because the brain growth spurt in both human and nonhuman primates extends over a much longer time period than in the rat, matching the timing of a developmental event between human and nonhuman primates is less problematic than matching the same between primates and rodents. Most investigators are convinced that anesthesia-induced neurotoxicity occurs, especially in the setting of a brain made vulnerable by developmental or pathological processes or perhaps by genomic predisposition, but that causality is presently unclear and probably multifactorial. To minimize the risk to children from exposure to anesthetics, it is necessary to understand anesthetic-induced effects on the developing nervous system and to distinguish their effects from those due to the physiologic changes that can also occur during anesthesia (for example, alterations in brain metabolism). Thus, further research in the nonhuman primate is urgently needed to determine which agents and procedures incur the greatest risk for subsequent brain dysfunction--including cognitive deficits--for both the very young and the elderly. In addition the threshold doses and durations necessary for safe and effective anesthetic exposures, as well as possible protective strategies must be determined.

Evidence for and Characteristics of, Anesthetic-Induced Neuronal Cell Death in the Developing Monkey Brain

The first report regarding neuronal cell death in nonhuman primates exposed perinatallhy to anesthetics was published in 2007 [5]. This study focused on the representative general anesthetic, ketamine (a non-competitive NMDA receptor antagonist). Ketamine was administered as an intravenous infusion to perinatal rhesus monkeys at doses sufficient to produce a light surgical plane of anesthesia [5].

The neurotoxic effects of these ketamine exposures were examined several hours after the end of the infusions, based on the hypothesis that ketamine induces an up-regulation of the NMDA NR1 receptor subunit, causing neurons to be more vulnerable to the excitotoxic effects of endogenous glutamate after ketamine washout. Ketamine infusion for 24 hours was shown to produce a large increase in the number of caspase 3-positive neurons in layers II and III of the frontal cortex in postnatal day 5 (PND 5) monkey brains compared with controls. Caspase 3-positive neurons in these layers still maintained their typical pyramidal morphology and neuronal processes. Although a few caspase 3-positive neurons were observed in some additional brain areas including the hippocampus, thalamus, striatum and amygdala, no significant difference was detected between ketamine-treated and control monkeys in those areas. Ketamine infusion had no detectable adverse effects on neurons in the cerebellum.

The 24-hour ketamine infusions produced elevated neuronal cell death as indicated by the increased number of TUNEL-positive cells in PND 5 infants. Numerous darkly stained TUNEL-positive cells exhibiting the typical nuclear condensation and fragmentation indicative of enhanced apoptotic cell death were observed in ketamine-infused PND 5 monkeys (Figure 1, [5]).

The TUNEL assay relies on the detection of fragmented DNA strands. At the EM level, direct evidence of increased neuronal cell death in PND 5 monkeys treated with ketamine was confirmed by the observation of representative nuclear condensation and fragmentation (apoptosis) and necrotic characteristics, including neuronal mitochondrial and neuronal cell body swelling. In contrast, PND 5 control monkeys neurons exhibited an intact cytoplasm and nuclear membrane.

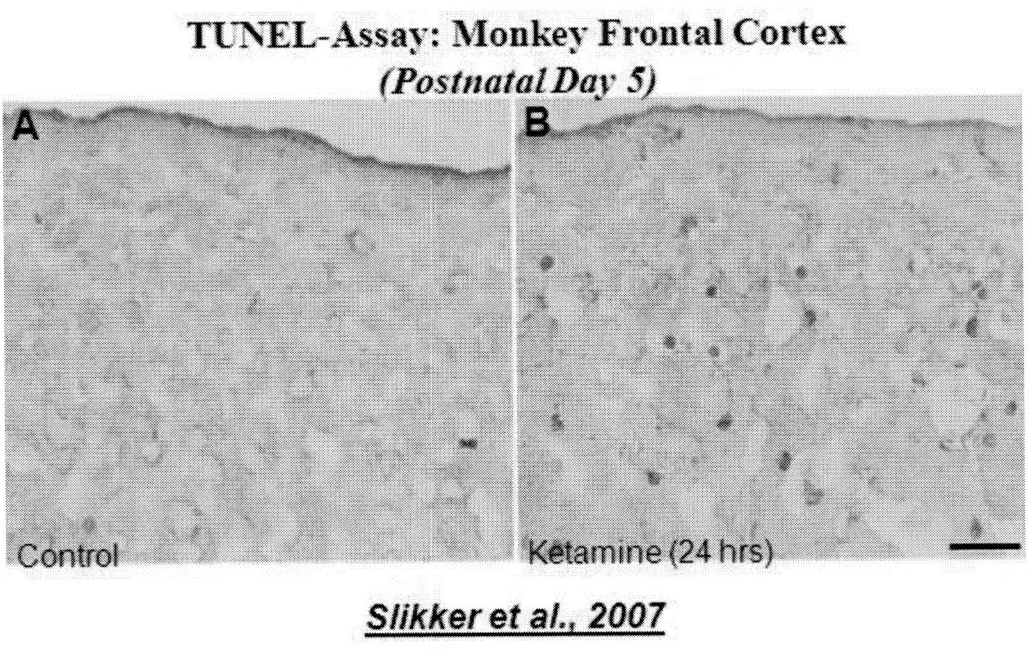

Figure 1. Ketamine-induced neurodegeneration assessed by TUNEL labeling in the frontal cortex of PND 5 monkeys. Representative photographs indicate that TUNEL-positive cells are more numerous in layers II and III of the frontal cortex in 24-h ketamine-infused PND 5 monkeys (B). Only a few TUNEL-positive cells were observed in the PND 5 controls (A). Scale bar = 60 μm.

It should be noted that the pattern or topography of ketamine-induced neurodegeneration in developing monkeys is very different from that reported in developing rodents. Our previous *in vivo* monkey study [5] demonstrated that ketamine-induced cell death is both apoptotic and necrotic in nature and restricted primarily to cortical layers II and III. In previous studies in the 7 day old rat, however, significant increases in apoptosis only were seen in multiple brain regions, particularly in the thalamus, after ketamine administration [2, 4, 6]. Also, the developmental lesions caused by ketamine in the young rat are distinctly different from those that occur in adult rats, which are characterized by vacuole formation in the retrosplenial cortex [7]. However, the consequences of altered apoptotic events that occur during development may be far more serious than the lesions observed in adult animals because a larger proportion of brain neurons may be compromised and the entire life of the animal can be affected.

Consistent with the *in vivo* data, our group [8] first used rhesus monkey frontal cortical cells in culture to determine the robustness of ketamine-induced developmental neurotoxicity in nonhuman primate tissue. Observations of increased internucleosomal DNA fragments (mostly apoptotic) and significant increases in LDH release (mostly necrotic), coupled with decreased mitochondrial MTT metabolism (reduction of total cell viability) suggested that ketamine-induced cell death in monkey cultures was characterized by both apoptosis and necrosis. We also confirmed the previous findings in rats [9] that ketamine administration during the perinatal period produces a dose-dependent increase in neurotoxicity; exposure of developing monkey brain cells to ketamine at concentrations as high as 10 or 20 μM caused both apoptosis and necrosis. The mechanism by which apoptosis was induced is hypothesized to be associated with a calcium overload via glutamatergic stimulation of compensatorily up-regulated NMDA receptors that exceeds the buffering capacity of mitochondria and leads to the increased generation of reactive oxygen species. An increase in the Bax/actin ratio and NF-κB were also found to be associated with ketamine induced cell death [8].

Relationships between Anesthetic-Induced Neurotoxicity and Developmental Stage at Time of Exposure

In addition to PND 5 monkeys, ketamine-induced neuronal degeneration was assessed in gestational day (GD) 122 and PND 35 monkeys [5]. Similar to PND 5 monkeys, the GD 122 monkey fetuses also showed clear ketamine-induced neuronal cell damage, whereas PND 35 monkeys did not (Figure 2, [6]). GD 122 monkey fetuses and PND 5 monkeys, thus, are sensitive to ketamine-induced cell death whereas PND 35 monkeys are not (less synaptogenesis is occurring at this age). Although a complete understanding of neuronal cell sensitivity to ketamine in the primate is not possible from these few early studies, it is apparent that rhesus monkeys are sensitive during the last 25% of gestation (GD 122) to sometime before PND 35. Equating relative stages of development between human and animal models is critical for the extrapolation of safety assessment data. It is generally believed that monkey fetuses and infants, and humans are more similar with respect to stage of maturation at birth as compared to rats that are relatively immature at birth. For example, both humans and rhesus monkeys are born with their eyes open at birth, whereas rat pups are not. Thus, PND 7 rat pups are more similar in maturation to monkeys or humans late in gestation than they are to infants. According to a recent review [10], the GD 123 monkey fetus is roughly equivalent to the GD 199 human fetus as determined by cortical development, and both are in the range of 75–80% of normal term. Also, NMDA receptor binding sites are present in the human fetal brain by GD 115, increase until GD 140–150, and then decrease slightly by GD 168–182 [11]. The localization of NMDA receptors in monkey cortex is similar to that seen in humans [12].

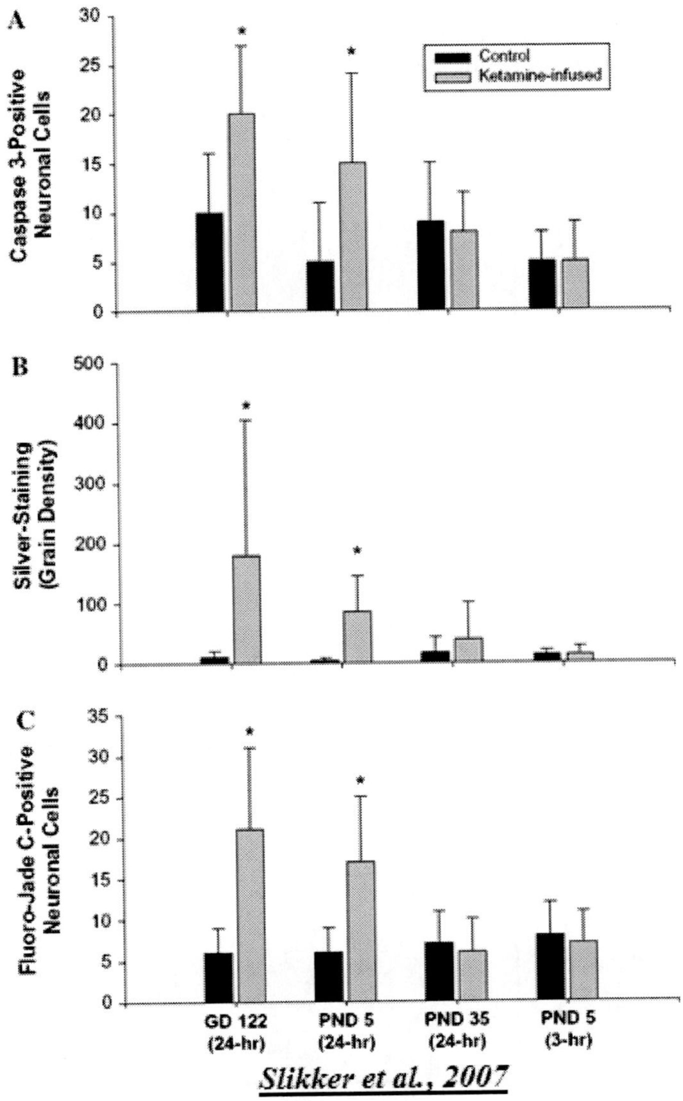

Figure 2. Quantitative analyses of ketamine-induced neurodegeneration in PND 5 monkeys after different anesthetic exposure durations. A significant enhancement of neurotoxicity was detected in animals exposed for 9 or 24 h as indicated by caspase-3- (A), Fluoro-Jade C- (B) and silver- (C) staining data (presented as means ± SD). *Indicates ketamine-treatment was statistically different from controls ($P < 0.05$) using one-way ANOVA. No significant neurotoxic effects were observed when the animals were exposed to ketamine for only 3 h and the neurodegeneration observed after a 24-h exposure was more intense than that observed after a 9-h exposure. For each condition, at least three animals were used.

Advances in pediatric and obstetric surgery have resulted in an increase in the complexity, duration and number of anesthetic procedures practiced. For example, pediatric heart transplant surgeries are more common now than earlier and theyy can take as long as 12 hours, depending on the patient's medical condition, previous surgeries and other factors. To minimize risks to children resulting from long exposures to anesthetics, it is important to determine if there are any adverse effects on CNS structure/function in pediatric populations and, if so, what factors contribute to such effects. One of the primary goals of anesthetic studies in the developing monkey was to determine whether there is an anesthetic duration below which no significant ketamine-induced neuronal cell death can be detected.

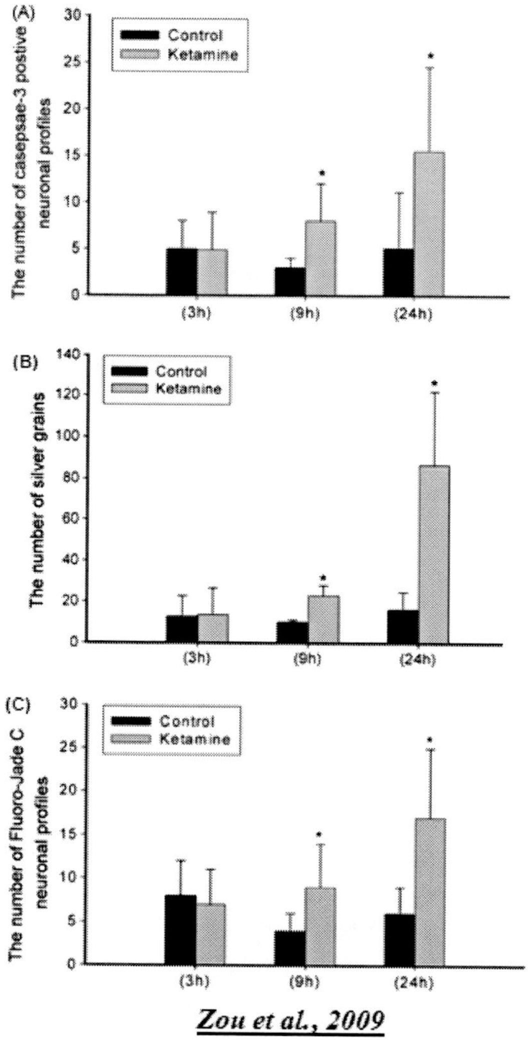

Figure 3. Quantitative analyses of ketamine-induced neurodegeneration at different developmental stages. (A): Quantitation of caspase-3 positive neurons; (B): Densitometric analysis of silver stain; (C): Quantitation of Fluoro-Jade C positive stain. For each condition, three animals were randomly assigned to treatment and control groups (N = 3/group). Data are presented as means ± SD. *A probability of $P < 0.05$ was considered significant (one-way ANOVA). While GD 122 and PND 5 monkeys demonstrated significant increases in neurodegenerative cells after a 24-h ketamine infusion, PND 35 monkeys were not different from control monkeys.

To address this issue, the neurotoxic effects of ketamine were explored in PND 5 monkeys anesthetized with ketamine for 3, 9 or 24 hours [6]. Significantly increased numbers of silver-impregnated and Fluoro-Jade C-positive degenerating neurons in layers II and III of the frontal cortex were observed after 9 and 24-h ketamine infusions compared to controls. The extent of neurodegeneration seen after a 24-h exposure was greater than that seen after a 9-h exposure, while a 3-h ketamine infusion did not have such effects, indicating that the amount of ketamine-induced neuronal cell death is dependent on the duration of exposure (Figure 3, [5]).

Physiological Parameters and Pharmacodynamic Outcomes Associated with Anesthetic-Induced Developmental Neurotoxicity

The effects of ketamine anesthesia and restraint on cardiovascular parameters in adult monkeys are well known [13-15], however, there was little information available for infants and pregnant monkeys. Also, little information was available about the effects of long-term ketamine infusion. To understand the effect of anesthetics on brain development, the effects of the anesthetic chemical must be distinguished from those of the physiological changes that occur during anesthesia. In our anesthetic studies in the developing monkey, the procedures followed for the maintenance and monitoring of experimental subjects during anesthesia have been previously described in detail [5]. All monkeys tolerated the procedures well and recovered from anesthesia uneventfully. Some of the physiological parameters changed during the 24-h experimental period in the ketamine-treated group; however, physiological values such as body temperature, blood glucose, and O_2 saturation remained within normal ranges for both control and ketamine-treated animals. For example, the percent oxygen saturation averaged 94% or above for all groups, while the heart and respiration rates were lower in ketamine-treated monkeys than in the corresponding (non-anesthetized) controls. Expired CO_2 concentrations were higher in treated than in control animals, and the changes were similar for PND 5 and PND 35 monkeys. Ketamine-treated pregnant female monkeys (GD 122) had lower blood pressure than controls [5]. It is noteworthy that ketamine is used clinically to provide anesthesia because it lacks the cardiorespiratory depression seen with most other general anesthetic agents [16] and its sympathomimetic properties counteract the cardio-depressive properties of other agents such as propofol [17]. In clinical practice, high plasma and brain concentrations of ketamine result in dissociative anesthesia, amnesia, a rise in arterial pressure, increased heart rate and cardiac output, and elevated intracranial pressure with relative preservation of airway reflexes and respiration. The abnormally high heart rate and blood pressure observed in the control monkey subjects may have been due to the transient restraint stress associated with obtaining these measurements. Body temperature and blood glucose were not different between control and treated monkeys. These data indicate that, as previously reported [5], ketamine-induced neurotoxicity was not due to hypothermia, hypoxia or hypoglycemia.

Plasma ketamine concentrations are critical exposure parameters associated with neuronal cell death in animals. In perinatal monkeys, steady-state plasma ketamine concentrations were achieved during 6–12 h of anesthesia. These levels in monkeys (10-25 μg/ml) are 5–10 times higher than those observed in humans (2–3 μg/ml) [5], but were the minimum necessary to maintain anesthesia in this animal model. Monkeys at various stages of development require different plasma ketamine levels to maintain an anesthetic plane. PND 35 monkeys required higher plasma ketamine concentrations to maintain the same plane of anesthesia compared to PND 5 animals. An important observation was that while the plasma concentration of ketamine needed to maintain anesthesia was the highest [5] in the older monkeys (PND 35), there was no evidence of increased neuronal cell death in those animals (Figure 2, [6]). In younger monkeys (PND 5), where neuronal cell loss was evident, the plasma levels averaged approximately 10 μg/ml, which is only 3 to 5 times higher than the plasma levels reported for humans [5].

How the Complex Behavioral Capabilities of the Rhesus Monkey Provide Opportunities to Study the Effects of Anesthetic Exposure during Development

Currently, the effects of specific anesthetic and surgical variables on subsequent cognitive performance are still not completely understood, nor have any molecular, cellular, or pathophysiological steps linking peri-operative events with cognitive outcomes been discerned from the human data. The choice and standardization of the cognitive domain to be tested, the timing of testing, controlling for anesthesia, surgery, inter-current medications, and co-morbidities all served to confound that data from human studies. The physiology and behavioral repertoire of the nonhuman primate and their similarity to those of humans make the monkey an excellent animal model for studying the effects of general anesthetic agents, including their effects on subsequent cognitive function.

For example, ketamine, like its close relative phencyclidine (PCP), is a dissociative anesthetic that acts primarily through blockade of N-methyl-D-aspartate (NMDA)-type glutamate receptors. Unlike PCP, ketamine is commonly used for a variety of pediatric procedures [18]. Concerns over the potential adverse effects of exposures to ketamine were piqued by the findings that blockade of NMDA receptors by ketamine and related compounds causes a robust increase in apoptotic cell death in the rat during the brain growth spurt period [2]. These findings were subsequently replicated and extended by others and in our own laboratories [4, 6, 19, 20], where it has been repeatedly demonstrated that multiple doses of ketamine given to neonatal rat pups on PND 7 (the peak of the brain growth spurt in rats) leads to a massive increase in apoptotic neuronal degeneration. Similar findings have also been demonstrated in mice [21]. To determine, in rhesus monkeys, if there are subsequent deficits in brain function associated with a single 24-h bout of ketamine-induced general anesthesia during the neonatal period, the National Center for Toxicological Research (NCTR) Operant Test Battery (OTB) [22, 23] was used to assess the brain function of these animals. The OTB contains several complex positively-reinforced tasks in which correct performance is thought to depend on relatively specific and important brain functions which include learning, color and position discrimination, motivation and short-term memory. The similarity in OTB performance between monkeys and children [23] is of particular importance with regard to extrapolating to humans the neurobehavioral (and possibly neurotoxic) effects of drugs and toxicants as determined in the monkey model. Additionally, the demonstration that several measures of OTB performance correlate highly with measures of intelligence in children [24] serves to highlight the relevance of such measures.

In an ongoing behavioral study [25], six monkeys were exposed on PND 5 or 6 to intravenous ketamine anesthesia to maintain a light surgical plane for 24 h and six control animals were unexposed. At 7 months of age all animals were weaned and began training to perform a series of cognitive function tasks as part of the National Center for Toxicological Research (NCTR) Operant Test Battery (OTB). The OTB tasks used here included those for assessing aspects of learning, motivation, color discrimination, and short-term memory. Subjects responded for banana-flavored food pellets by pressing response levers and press-plates during daily (M-F) test sessions (50 min) and were assigned training scores based upon their individual performance. As reported earlier [25] beginning around 10 months of age, control animals significantly outperformed (had higher training scores than) ketamine-exposed animals for approximately the next 10 months. For animals now over two and one-half years of age, the cognitive impairments continue to manifest in the ketamine-exposed group as poorer performance in the OTB learning and color and position discrimination tasks as deficits in accuracy of task performance and in response speed. There are also apparent differences in the motivation of these animals which may be impacting OTB performance. These observations demonstrate that a single 24-h episode of ketamine anesthesia, occurring during a sensitive period of brain development, results in very long-lasting deficits in brain function in primates and provides proof-of-concept that general anesthesia during critical periods of brain development can result in subsequent functional deficits.

The Use of a Nonhuman Primate Model to Decrease the Uncertainty in Extrapolating Pre-Clinical Data to the Human Condition

Evidence in support of a correlation between surgery and subsequent neuro-physiological changes has accumulated [26-28]. The use of a nonhuman primate model to decrease the uncertainty in extrapolating pre-clinical data to the human condition (e.g. peri-operative neurotoxicity) continues to garner considerable interest among anesthesiologists and toxicological researchers, with growing recognition to be anticipated from surgeons. Currently, a host of mechanistic studies have been completed or are underway which have been helpful in providing a rationale for the overall concern of peri-operative neurotoxicity, teasing apart the causalities, refining hypotheses, and suggesting clinical strategies to test for the problem in patients. These have ranged from cell culture to histopathology to animal behavioral studies – including the nonhuman primate [5, 25, 29, 30]. To date, data from rodents and nonhuman primates have demonstrated neurotoxic effects of anesthetic drugs on the developing brain that are associated with later deficits in brain functions including learning, the ability to perform simple visual discriminations, motivation and speed of psychomotor processing [25]. However, there are currently no clinical data providing evidence that the clinical use of anesthetics is associated with signs of developmental neurotoxicity or subsequent cognitive deficits in humans. It is essential to continue studies in monkeys to obtain valuable information on the time course and severity of observed deficits. It will also be necessary to determine whether injured brain tissue can recover with no or minimal loss of function, or whether injured brain tissue can be protected by the co-administration of anti-oxidant agents. Shorter durations of anesthesia cause less or no cell death in monkeys: whether exposures to anesthetics will cause cell death in humans is still unknown, but it is likely that shorter exposure durations will have less impact than longer durations. Although drug combinations are commonly used in pediatric surgical procedures, there is a huge data gap on the neurodegenerative effects associated with anesthetic drug combinations.

A growing body of data indicates that molecular imaging with isotope-labeled biomarkers (radio-tracers) may help to detect neurotoxicity in neonates, infants, young and adult monkeys and humans. The high-resolution positron emission tomography scanner (microPET) provides *in vivo* molecular imaging at a sufficient resolution to resolve both major structures and neuronal activities in the nonhuman primate brain. To determine whether prolonged pediatric anesthetic exposure is associated with subsequent long-term cognitive deficits, anesthetic-induced neuro-degeneration can be explored by monitoring changes in the uptake (binding) of a radiotracer (e.g., [^{18}F]-Benzodiazepine Receptor ligand (PBRS), a neurotoxicity- and gliosis-related biomarker) in specific regions of interest (ROIs) in the monkey brain. Meanwhile, Operant Test Battery tasks, including those for assessing aspects of learning, motivation, color and position discrimination, and memory can be useful tools when used in parallel with longitudinal microPET imaging assessments to delineate the time course of cognitive performance deficits and underlying biochemical changes.

It seems that most investigators are convinced that peri-operative neurotoxicity occurs, especially in the setting of a brain made vulnerable by developmental or pathological processes or perhaps by genomic predisposition, but causality is presently unclear and probably multifactorial. There are yet many questions to answer before the findings of anesthetic-induced neurotoxicity observed in animals can be related to effects in humans: however, the use of a nonhuman primate model combined with molecular imaging tools and dynamic behavioral tests might provide the most expeditious approach toward decreasing the uncertainty in extrapolating pre-clinical data to the human condition.

CONCLUSION

It has been proposed that anesthetic-induced neurotoxicity depends on the amount (dose) given, the duration of the exposure, the route of administration, the receptor subtype activated, and the stage of the development at the time of exposure. These factors are important because they can help identify thresholds of exposure for producing neurotoxicity in the developing nervous system. Data from rhesus monkeys suggest that anesthetic-induced neurodegeneration is exposure-duration and developmental-stage dependent. Shorter durations of anesthesia (e.g. a ketamine infusion for 3 hours) do not cause neuronal cell death even when administered during sensitive developmental stages in rhesus monkeys. The neuronal cell death induced by anesthetics in the primate is both apoptotic and necrotic in nature.

Further research (e.g. dynamic molecular imaging and brain function assessments in nonhuman primate models) is urgently needed to determine which developmental stage and procedures incur greater risk of subsequent cognitive effects, whether in the young or the elderly, and what steps may be taken to ameliorate the risk.

Disclaimer: This document has been reviewed in accordance with United States Food and Drug Administration (FDA) policy and approved for publication. Approval does not signify that the contents necessarily reflect the position or opinions of the FDA. The findings and conclusions in this report are those of the authors and do not necessarily represent the views of the FDA.

REFERENCES

[1] Machado CJ, Bachevalier J. Non-human primate models of childhood psychopathology: the promise and the limitations. *J. Child Psychol. Psychiatry*. 2003 Jan;44(1):64-87.

[2] Ikonomidou C, Bosch F, Miksa M, Bittigau P, Vockler J, Dikranian K, Tenkova TI, Stefovska V, Turski L, Olney JW. Blockade of NMDA receptors and apoptotic neurodegeneration in the developing brain. *Science*. 1999 Jan 1;283(5398):70-4.

[3] Jevtovic-Todorovic V, Hartman RE, Izumi Y, Benshoff ND, Dikranian K, Zorumski CF, Olney JW, Wozniak DF. Early exposure to common anesthetic agents causes widespread neurodegeneration in the developing rat brain and persistent learning deficits. *J. Neurosci*. 2003 Feb 1;23(3):876-82.

[4] Scallet AC, Schmued LC, Slikker W, Jr., Grunberg N, Faustino PJ, Davis H, Lester D, Pine PS, Sistare F, Hanig JP. Developmental neurotoxicity of ketamine: morphometric confirmation, exposure parameters, and multiple fluorescent labeling of apoptotic neurons. *Toxicol. Sci*. 2004 Oct;81(2):364-70.

[5] Slikker W, Jr., Zou X, Hotchkiss CE, Divine RL, Sadovova N, Twaddle NC, Doerge DR, Scallet AC, Patterson TA, Hanig JP, Paule MG, Wang, C. Ketamine-induced neuronal cell death in the perinatal rhesus monkey. *Toxicol Sci*. 2007 Jul;98(1):145-58.

[6] Zou X, Patterson TA, Divine RL, Sadovova N, Zhang X, Hanig JP, Paule MG, Slikker W, Jr., Wang C. Prolonged exposure to ketamine increases neurodegeneration in the developing monkey brain. *Int. J. Dev. Neurosci*. 2009 Nov;27(7):727-31.

[7] Olney JW. Neurotoxicity of NMDA receptor antagonists: an overview. *Psychopharmacol. Bull*. 1994;30(4):533-40.

[8] Wang C, Sadovova N, Hotchkiss C, Fu X, Scallet AC, Patterson TA, Hanig J, Paule MG, Slikker W, Jr.. Blockade of N-methyl-D-aspartate receptors by ketamine produces loss of postnatal day 3 monkey frontal cortical neurons in culture. *Toxicol. Sci*. 2006 May;91(1):192-201.

[9] Wang C, Sadovova N, Fu X, Schmued L, Scallet A, Hanig J, Slikker W, Jr.. The role of the N-methyl-D-aspartate receptor in ketamine-induced apoptosis in rat forebrain culture. *Neuroscience*. 2005;132(4):967-77.

[10] Clancy B, Finlay BL, Darlington RB, Anand KJ. Extrapolating brain development from experimental species to humans. *Neurotoxicology*. 2007 Sep;28(5):931-7.

[11] Haberny KA, Paule MG, Scallet AC, Sistare FD, Lester DS, Hanig JP, Slikker W, Jr. Ontogeny of the N-methyl-D-aspartate (NMDA) receptor system and susceptibility to neurotoxicity. *Toxicol. Sci*. 2002 Jul;68(1):9-17.

[12] Huntley GW, Vickers JC, Morrison JH. Quantitative localization of NMDAR1 receptor subunit immunoreactivity in inferotemporal and prefrontal association cortices of monkey and human. *Brain Res*. 1997 Feb 28;749(2):245-62.

[13] Bush M, Custer R, Smeller J, Bush LM. Physiologic measures of nonhuman primates during physical restraint and chemical immobilization. *J. Am. Vet. Med. Assoc.* 1977 Nov 1;171(9):866-9.

[14] Golub MS, Anderson JH. Adaptation of pregnant rhesus monkeys to short-term chair restraint. *Lab. Anim. Sci.* 1986 Oct;36(5):507-11.

[15] Mori Y, Petersen B, Enderle N, Congdon WC, Baker B, Meyer S. Effect of ketamine on cardiovascular parameters and body temperature in cynomolgus monkeys. *J. Amer. Assoc. Lab. Anim. Sci.* 2006;45:128.

[16] Zielmann S, Kazmaier S, Schnull S, Weyland A. [S-(+)-Ketamine and circulation]. *Anaesthesist.* 1997 Mar;46 Suppl 1:S43-6.

[17] Badrinath S, Avramov MN, Shadrick M, Witt TR, Ivankovich AD. The use of a ketamine-propofol combination during monitored anesthesia care. *Anesth. Analg.* 2000 Apr;90(4):858-62.

[18] Kohrs R, Durieux ME. Ketamine: teaching an old drug new tricks. *Anesth. Analg.* 1998 Nov;87(5):1186-93.

[19] Hayashi H, Dikkes P, Soriano SG. Repeated administration of ketamine may lead to neuronal degeneration in the developing rat brain. *Paediatr. Anaesth.* 2002 Nov;12(9):770-4.

[20] Shi Q, Guo L, Patterson TA, Dial S, Li Q, Sadovova N, et al. Gene expression profiling in the developing rat brain exposed to ketamine. *Neuroscience.* 2010 Mar 31;166(3):852-63.

[21] Young C, Jevtovic-Todorovic V, Qin YQ, Tenkova T, Wang H, Labruyere J, Olney, J.W. Potential of ketamine and midazolam, individually or in combination, to induce apoptotic neurodegeneration in the infant mouse brain. *Br. J. Pharmacol.* 2005 Sep;146(2):189-97.

[22] Paule MG. Methods of Behavioral Analysis in Neuroscience. Buccafusco, J.J, Ed: CRC Press LLC, Boca Raton, FL; 2001.

[23] Paule MG, Cranmer JM. Complex brain function in children as measured in the NCTR monkey operant test battery. Advances in Neurobehavioral Toxicology: Applications in Environmental and Occupational Health. B.L. Johnson, Ed: Lewis Publishers, Chelsea, MI; 1990.

[24] Paule MG, Meck WH, McMillan DE, McClure GY, Bateson M, Popke EJ, Chelonis JJ, Hinton SC. The use of timing behaviors in animals and humans to detect drug and/or toxicant effects. *Neurotoxicol. Teratol.* 1999 Sep-Oct;21(5):491-502.

[25] Paule MG, Allen RR, Liu F, Zou X, Hotchkiss C, Hanig J, Patterson TA, Slikker W, Wang C. Ketamine anesthesia during the first week of life can cause long-lasting cognitive deficits in rhesus monkeys. *Neurotoxicology and Teratology*. Submitted.

[26] Biedler A, Juckenhofel S, Larsen R, Radtke F, Stotz A, Warmann J, Braune E, Dyttkowitz A, Henning F, Strickmann B, Lauven PM. [Postoperative cognition disorders in elderly patients. The results of the "International Study of Postoperative Cognitive Dysfunction" ISPOCD 1)]. *Anaesthesist.* 1999 Dec;48(12):884-95.

[27] Johnson T, Monk T, Rasmussen LS, Abildstrom H, Houx P, Korttila K, Kuipers HM, Hanning CD, Siersma VD, Kristensen D, Canet J, Ibanaz MT, Moller JT. Postoperative cognitive dysfunction in middle-aged patients. *Anesthesiology.* 2002 Jun;96(6):1351-7.

[28] Canet J, Raeder J, Rasmussen LS, Enlund M, Kuipers HM, Hanning CD, Jolles J, Korttila K, Siersma VD, Dodds C, Abildstrom H, Sneyd JR, Vila P, Johnson T, Munoz

Corsini L, Silverstein JH, Nielsen IK, Moller JT. Cognitive dysfunction after minor surgery in the elderly. *Acta Anaesthesiol. Scand.* 2003 Nov;47(10):1204-10.
[29] Xie Z, Tanzi RE. Alzheimer's disease and post-operative cognitive dysfunction. *Exp. Gerontol.* 2006 Apr;41(4):346-59.
[30] Rizzi S, Ori C, Jevtovic-Todorovic V. Timing versus duration: determinants of anesthesia-induced developmental apoptosis in the young mammalian brain. *Ann. N.Y. Acad. Sci.* Jun;1199:43-51.

In: Monkeys: Biology, Behavior and Disorders
Editor: Rachel M. Williams, pp 181-197
ISBN: 978-1-61209-911-8

Chapter 9

VISUAL PROCESSING IN THE MONKEY

Benjamin S. Lankow and W. Martin Usrey*
Center for Neuroscience and the Departments of Neurobiology, Physiology & Behavior and Neurology, University of California, Davis

INTRODUCTION

The monkey visual system begins with the eyes and includes multiple brain areas and processing streams. Consequently, more of the monkey brain is devoted to vision than to any other sensory modality. Like other sensory systems, the visual system is organized hierarchically both in structure and function, and subsequent levels of the hierarchy are more specialized in their processing of visual features than their predecessors. Visual information is translated from the physical world into a neural code in the retina. This information, contained in the spiking activity of retinal ganglion cells, is then transmitted to the lateral geniculate nucleus (LGN) of the thalamus, a processing station that subsequently supplies the primary visual cortex (V1) with the majority of its input. From V1, visual information is routed into multiple branching streams that encompass more than 20 cortical areas, spanning regions whose response properties range from simple edge detection to complex representations of familiar conspecifics. The study of monkey vision thus requires a vast array of strategies and techniques, from the biochemical analysis of phototransduction in the retina, to computational modeling of neural networks that are capable of the processing feats required to construct a neural representation of our dynamically changing visual world. While the enormity of vision science can be daunting, we present in this chapter a basic foundation necessary for understanding how visual information is encoded and processed in the monkey nervous system, giving particular emphasis to the inseparable relationship between anatomy and function. This chapter will emphasize visual processing in the macaque monkey—an old world monkey that has been studied more comprehensively than other primates and serves as a model for understanding primate vision.

* Correspondence: Center for Neuroscience, University of California, Davis, 1544 Newton Court, Davis, CA 95618, 530 754-5468 ph, 530 757-8827 fax, wmusrey@ucdavis.edu

The Monkey Eye and Retina

Inputs: The Visual World

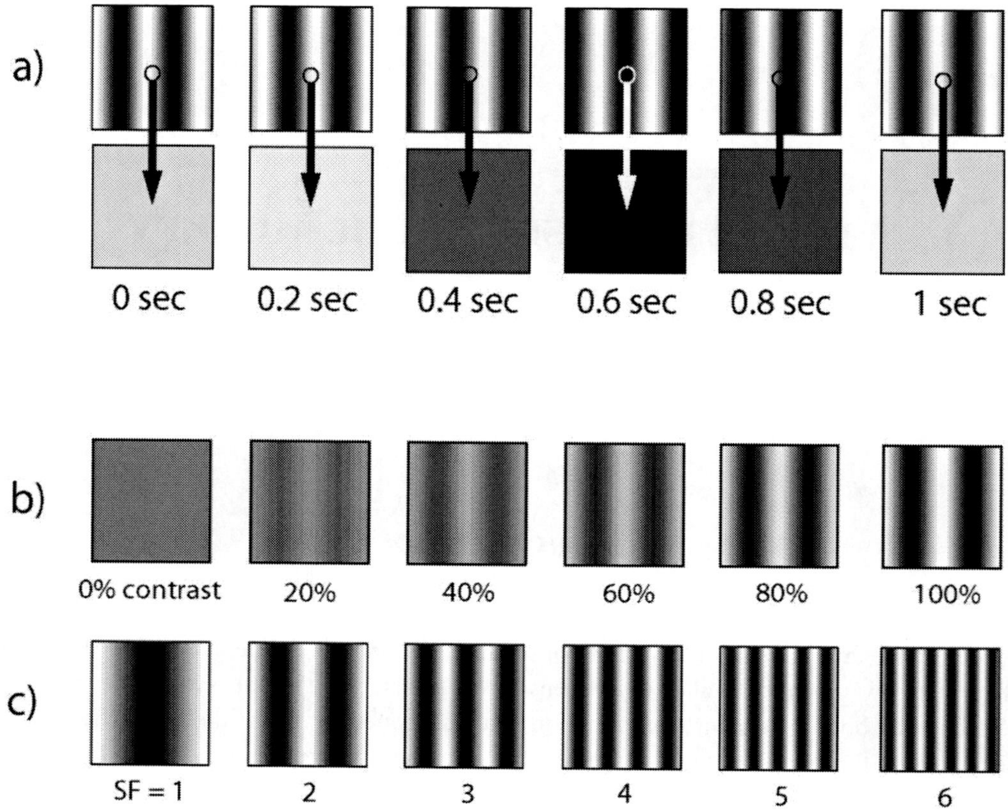

Figure 1. Sine-wave gratings can be used to study neuronal responses to temporal contrast, luminance contrast, and spatial contrast. a) A drifting sine-wave grating produces temporal contrast as a function of the temporal frequency (drift rate) of the grating. Higher or lower temporal contrast can be created by adjusting the drift rate of the grating. This grating has a spatial frequency of 2 cycles/ frame and a temporal frequency of 1 cycle/sec. b) Sine-wave gratings demonstrating luminance contrast as a function of sine-wave amplitude. c) Sine-wave gratings demonstrating spatial contrast by modulating the spatial frequency of the sine-wave grating.

The visual system has evolved to provide an animal with accurate information about the visual environment; it is useful, then, to spend some time at the outset of this chapter discussing what the visual environment consists of and what it is that the visual system measures. At the most basic level, the visual environment consists of a light source, referred to as an *illuminant*, and objects that reflect this light, referred to as *reflectants*. Different illuminants, such as the sun or a pair of headlights, give off different combinations of colors, or wavelengths, of light; the amount of energy that a particular light source emits at various wavelengths is described by the *spectral power distribution* of the illuminant. The sun emits a very broad spectrum of light, so as you might imagine, different objects in the environment reflect this light in quite different ways. Most objects or materials are biased toward reflecting certain wavelengths of light better than others-- for example, a rose petal that appears red reflects long-wavelength light very well; the foliage of a tree that appears green reflects medium-wavelength light very well. We refer to the object's ability to reflect different wavelengths of light as its *surface reflectance*. In a natural landscape, there is a rich array of reflectants redirecting the light emitted from the sun, moon, or stars; the visual system must utilize this reflected light to infer the boundaries, shape, and identity of objects, and therefore is adapted to measure *spatial contrast,* the fractional difference in luminance or color within regions of visual space. Likewise, the visual system is adapted to rapidly detect *temporal contrast*, the change in luminance of a particular region over time. Temporal contrast can be caused by the ongoing movement of the eyes, by the motion of objects relative to the environment, and by the relative motion of the visual environment as an animal navigates its habitat. Researchers studying vision typically use stimuli in the laboratory that can precisely reproduce desired aspects of the natural visual environment (reviewed in Wandell, 1995), such as spatial and temporal contrast; much of the information contained in this chapter has been discovered by probing the visual system with these reduced stimuli (figure 1) and precisely measuring the output of cells in response to the modulation of a chosen aspect of the stimulus.

Optics and Image Formation

The visual system is able to make use of the reflected light in the visual environment because of the ability of the optics of the eye to accurately focus an image onto the cells that convert light energy into neuronal responses. Light from the environment is refracted by the *cornea* and the *lens* so that it is in focus at the *retina* (figure 2). The cornea provides the majority of the refractive power of the eye due to the relative refractive indices of the cornea and that of air; the cornea, however, has a fixed refractive power, and it is the adjustable lens that is responsible for bringing objects in the visual environment into focus on the retina. The process of adjusting the refractive power of the lens is referred to as *accommodation*. The lens is connected via *zonule fibers* to the *ciliary muscle*, which forms a ring around the lens and decreases in diameter when it contracts. The lens is elastic, and when allowed to relax completely, is almost spherical. Thus, when the ciliary muscle contracts, the tension on the zonule fibers lessens and the lens becomes more spherical, increasing its refractive power. Conversely, when the ciliary muscle relaxes, its own elasticity pulls outwardly on the zonule fibers, flattening the lens and decreasing its refractive power.

The *pupil*, which is the aperture of the iris, controls the amount of light that enters the eye, and along with the cornea and lens, contributes to the clear focus of objects onto the retina. A balance must be struck in order to most effectively adjust the diameter of the pupil; under daylight conditions, the pupil can constrict to become quite small, and behaves much like the pinhole of a camera-- it narrows the angle of light rays that are capable of passing into the eye, and thus reduces aberrations in the image; too small of a pupil, however, would cause diffraction of the rays passing through the aperture and therefore cause blurring. Under low lighting conditions, the pupil must dilate to allow enough light into the eye to drive reliable visual responses from the retinal photoreceptors. Because the quality of the image will decrease as the diameter of the aperture increases, this illustrates the common theme of compromise in the workings of the visual system to allow for satisfactory function under a huge array of different conditions.

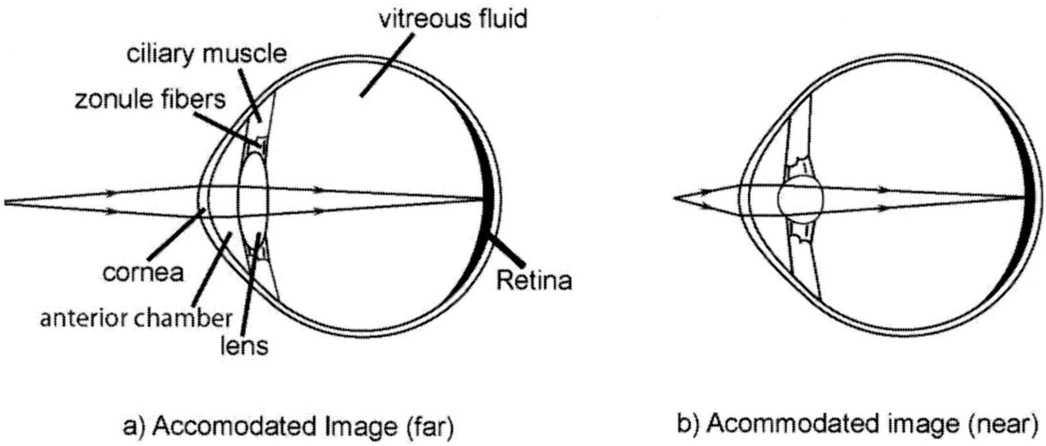

Figure 2. Schematic diagram of the optics of the eye. a) Relaxation of the ciliary muscle results in flattening of the lens, thereby reducing its refractive power; in this example, a distant object is brought into focus on the retina. b) Contraction of the ciliary muscle results in a more curved lens, thereby increasing its refractive power, and in this case bringing a near object into focus at the retina.

Gross Structure of the Retina: Cell Types, Layers, and Functional Organization

The retina is responsible for transforming information from the visual environment into neural activity that will be transmitted to the other visual areas of the brain. The retina is a part of the central nervous system, and develops from the same embryonic tissue as the brain. There are 5 classes of cells that compose the retina, and these function in either the detection (photoreceptors), processing (horizontal, bipolar, and amacrine cells), or transmission (retinal ganglion cells) of visual information. The organization of these cell types is highly conserved, and they form five characteristic layers that define the retina (figure 3), which is approximately 0.5 mm thick. Three of these layers are composed of the cell bodies, and two layers are composed of the dendrites and processes of the cells. The outer nuclear layer, furthest from the center of the eye, is composed of the cell bodies of the photoreceptors. The inner nuclear layer, the middle cell body layer, is composed of the cell bodies of the horizontal, bipolar, and amacrine cells. The innermost layer of retinal cells, closest to the center of the eye, is the ganglion cell layer, which is composed of the cell bodies of the output cells of the retina, which ultimately transmit visual information to the rest of the brain. The majority of the spatial processing of visual information in the retina occurs primarily through the interactions between these cell types; these interactions occur in the layers of processes between the cell body layers. The outer plexiform layer contains the processes of the photoreceptors, the bipolar cells, and the horizontal cells; the inner plexiform layer contains the processes of the retinal ganglion cells, amacrine cells, and bipolar cells.

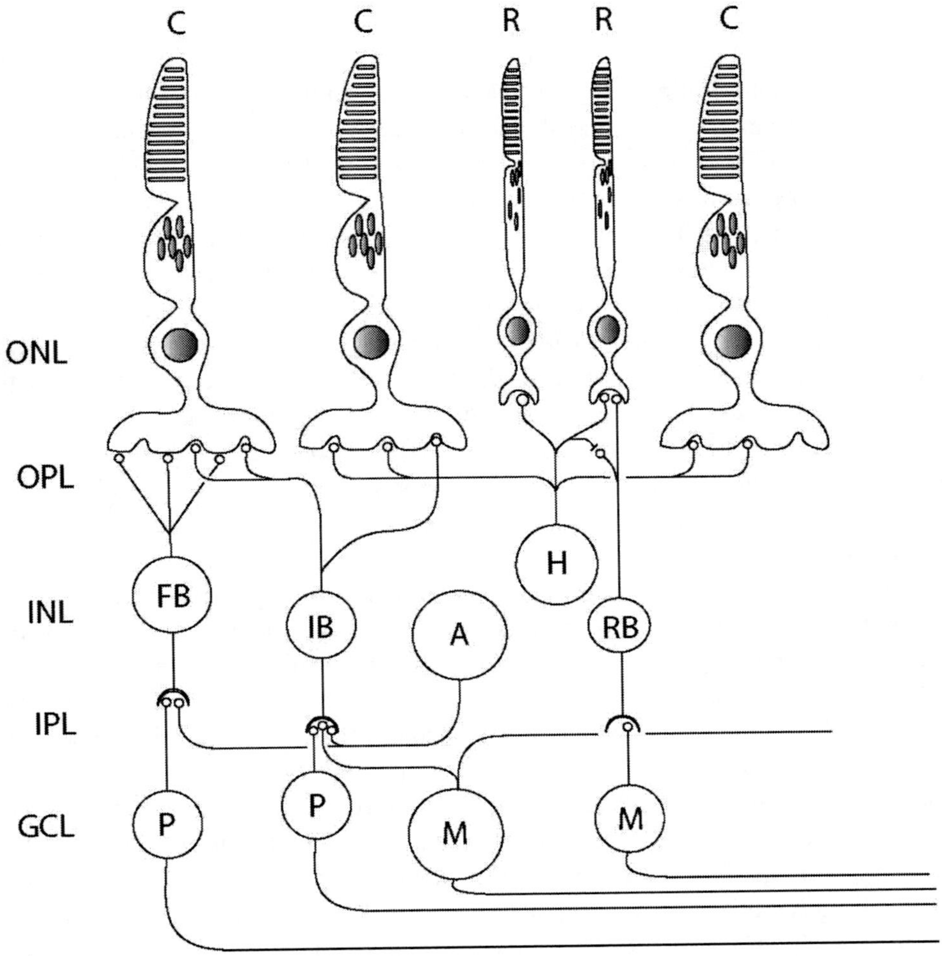

Figure 3. Layers and major cell types of the retina. There are 5 major classes of cell types and 5 layers in the retina. R: rod photoreceptor, C: cone photoreceptor, FB: flat bipolar cell, IB: invaginating bipolar cell, A: amacrine cell, H: horizontal cell, RB: rod bipolar cell, P: P-projecting retinal ganglion cell, M: M-projecting retinal ganglion cell, ONL: outer nuclear layer, OPL: outer plexiform layer, INL: inner nuclear layer, IPL: inner plexiform layer, GCL: ganglion cell layer.

Photoreceptor Properties and Distributions

Photoreceptors are responsible for the fast detection of photons of light that have been focused onto the retina by the optics of the eye. Different classes of photoreceptors are specialized for the accurate detection of photons under different lighting conditions. The cone photoreceptors are specialized for high resolution color vision in daylight (photopic) conditions, whereas the rod photoreceptors are specialized for vision under low luminance (scotopic) conditions. Both classes of photoreceptors, though morphologically discernible, have a similar gross organization: an outer segment in which phototransduction occurs, an inner segment containing mitochondria that provide energy, a cell body containing the nucleus, and an output structure that releases glutamate, an excitatory neurotransmitter (figure

3). The membrane voltage of the photoreceptor, which determines the graded release of glutamate, is controlled by ion channels in the membrane that are opened (gated) by cGMP.

Phototransduction functions through a pathway that controls the amount of cGMP present in the cell, thereby affecting the membrane potential and ultimately regulating transmitter release. The process occurs through a biochemical cascade that begins when photons of light are absorbed by a photopigment in the outer segment of the receptor. These photopigments are composed of a light-absorbing component, *11*-cis retinal, and one of several opsins expressed in the retina. The spectral sensitivities of the cone photoreceptors are determined by the light absorption properties of the particular opsin that it exclusively expresses. When a photon is absorbed, the G-protein *transducin* is activated, which subsequently activates a phosphodiesterase that hydrolyzes cGMP; the net effect of this cascade is the reduction of cGMP concentration inside the cell, causing cGMP-gated ion channels in the membrane to close, hyperpolarizing the cell. Thus the response of a photoreceptor cell to increases in light is to hyperpolarize and reduce the amount of neurotransmitter released. The converse is true in response to decrements in light, where the cell depolarizes and increases the release of neurotransmitter.

In humans and old-world primates, there are three cone types in the retina, which are defined by the particular opsin expressed, and hence the spectral sensitivity exhibited. Each cone type expresses only one form of the photopigment opsin, and is therefore classified as either L, M, or S (for long, medium, and short wavelength sensitivity). Because each cone classes responds to a broad distribution of wavelengths, the L, M, and S designations are not an indication of exclusive sensitivity, but rather the wavelength of light that is *optimally* absorbed by each cone class. Rod photoreceptors exclusively express the photopigment rhodopsin, and are optimally sensitive to light with a wavelength between that optimal for M and S cones. The distribution and tiling of photoreceptor types is not constant across the retina, but changes predictably with retinal eccentricity. In many diurnal primates, the fovea is composed entirely of cones. Rods, which are responsible for signaling under scotopic conditions, are densely packed in the peripheral retina, and dramatically fall off in number near the fovea. The absence of rods from the fovea is the reason that objects can be discerned more easily in the visual periphery at night. In nocturnal primates, such as owl monkeys, the photoreceptor array is relatively enriched in rods, and the retina lacks a fovea.

Receptive Field Construction: Retinal Circuitry at Work

The field of vision science is dominated by the concept of the *receptive field.* Receptive fields were originally defined as the region of the retina that must be illuminated in order to evoke a response from a cell (Hartline, 1938). This definition has evolved over the years and has been extended to include patterns, color, motion, and even more complex stimulus characteristics. At the level of a photoreceptor, the receptive field consists merely of photons of appropriate wavelength landing on the particular region of the retina in which the receptor is located. At the highest levels of the visual hierarchy, it can consist of complex motion patterns or specific faces. The circuitry of the retina is responsible for transforming the photoreceptor inputs into the more complex receptive fields of the retinal ganglion cells, which are capable of signaling spatial and temporal contrast, as well as motion in some cases. The first receptive field transformation that occurs after phototransduction is the structuring of *on* and *off* responses in the bipolar cells (figure 3). Each bipolar cell can be classified as either on or off, meaning it is excited by either increments or decrements of light. This response property is a consequence of the type of glutamate receptor that is expressed by the bipolar cell. Since all rod and cone photoreceptors exclusively release glutamate as a neurotransmitter, it is necessary that there is a mechanism to generate different responses to the same chemical; the effect that glutamate will have on the postsynaptic bipolar cell is determined by the presence of either excitatory ionotropic glutamate receptors, or inhibitory metabotropic receptors. These different classes of bipolar cells can be identified by the shape of the synapses they form with the photoreceptors; on-bipolar cells synapse within invaginations on the processes of photoreceptor cells, and off-bipolar cells synapse on outer flat portions of photoreceptor processes, leading to the naming convention of invaginating bipolars and flat bipolars, respectively. The bipolar cells synapse onto retinal ganglion cells, the sole class of output cells in the retina, and confer the on or off response properties of the ganglion cells (figure 4). Most ganglion cells receive synapses from a single class of bipolar cells, and therefore have either on or off response properties, although some ganglion cells receive synapses from both on and off bipolar cells, responding to both light increments and light decrements; these are known as on-off ganglion cells. While the retinal responses to changes in illumination over time are formed primarily by the feedforward connections of photoreceptors, bipolar cells, and ganglion cells, the spatial response properties are primarily mediated by lateral connections between these feedforward circuits. This occurs via the horizontal cells and the amacrine cells, which form synapses among locally residing feedforward cells. The horizontal cells are GABAergic, but the amacrine cell types utilize a wide array of neurotransmitters, particularly GABA, glycine, and acetylcholine. In the outer plexiform layer, the processes of the horizontal cells form synapses that influence the interactions between the photoreceptors and bipolar cells. These processes project laterally across the photoreceptors and bipolar cells, and connect on and off bipolar cells in a specific manner that transforms the spatial response properties of these cells from that of the photoreceptors into an antagonistic center-surround organization. These bipolar cell receptive fields can be characterized as either on-center/off surround or off center/on surround. This center/surround organization is subsequently passed on to the retinal ganglion cells (figure 4). Our understanding of the role of the amacrine cells in the functional circuitry of the retina is murkier than that of the horizontal cells; like the horizontal cells, amacrine cells send processes laterally (in the inner plexiform layer) across the retina, and form synapses between feedforward cells, in this case the bipolar cells and ganglion cells.

Retinal Output: Parallel Pathways and Their Targets

Counterintuitively, the innermost cell layer of the retina is composed of the output cells, the retinal ganglion cells. The axons of retinal ganglion cells traverse the retina and exit as a condensed collection of fibers, forming a cone of axons passing through the retinal layers on their way out, called the *optic disk*. The location of the optic disk is conserved, and due to the displacement of photoreceptors from this region, it constitutes a "blind spot" in the visual field. The outer surface of the optic disk is the formally defined start of the *optic nerve*, which projects primarily to the lateral geniculate nucleus (LGN) of the thalamus and to the superior colliculus. The retinal ganglion cells projecting to the LGN can be grouped into three major functional classes, each with its own distribution of response properties, morphology, and projection pattern; the optic nerve, therefore, carries three major channels of information destined for the cortex.

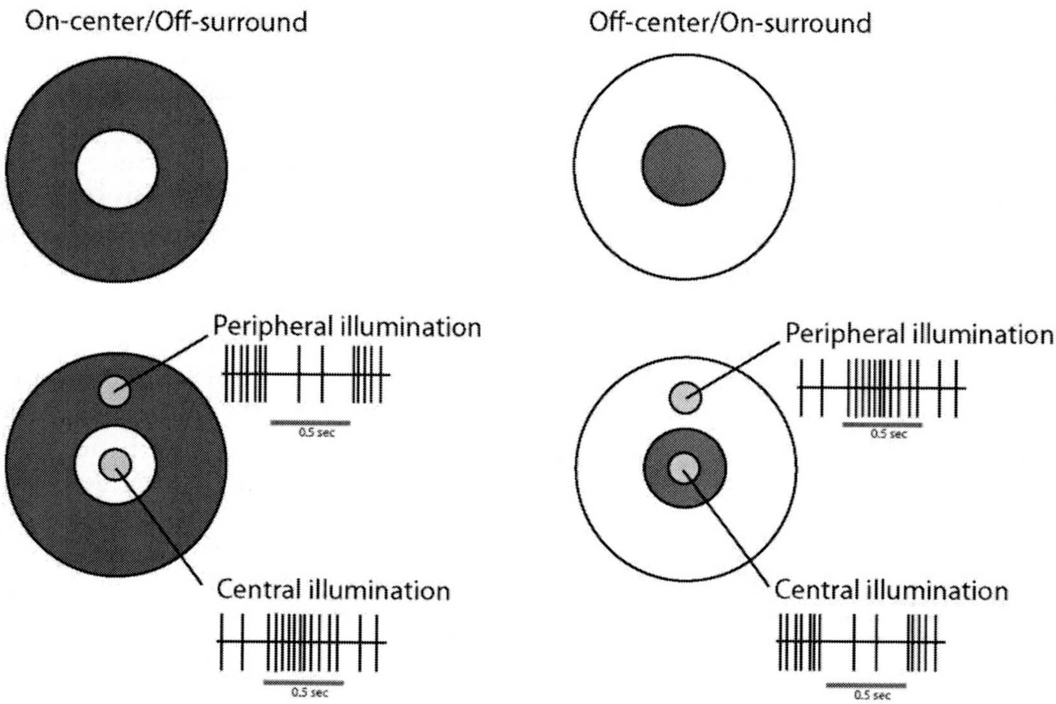

Figure 4. Response properties of on-center and off-center retinal ganglion cells. Light gray circles indicate regions of the receptive field that have been illuminated to yield the corresponding responses; the duration of illumination is indicated by the gray bars under the responses. Vertical lines indicate action potentials.

The P Channel

The P cells project to the parvocellular (small cell) layers of the LGN, and compose the majority of the ganglion cells in the retina (approximately 1,000,000); because of their small size, they are also referred to as *midget* ganglion cells, although they will be referred to exclusively as P cells in this chapter to avoid confusion with the M/magnocellular cells. P cells have small dendritic arbors indicative of the very narrow convergence they receive from bipolar cells, and consequently have small receptive fields. In the fovea, P cells receive input from a single bipolar cell, which receives input from a single photoreceptor, thereby strongly delimiting the area of visual space for which the P cell will carry information. At further retinal eccentricities, the convergence expands, so that bipolar cells receive input from several cone photoreceptors, and several bipolar cells connect to a single P cell. The effect of this convergence is an increase in the size of the receptive field—that is, because the P cell receives inputs from a larger area of the retina, the region of visual space that it is responsive to increases.

In old-world monkeys, the input to P cells originates from the L and M cones, and these photoreceptors respond to a *distribution* of wavelengths centered around a preferred value. To distinguish specific colors, therefore, there must be a mechanism for the comparison of the responses of different cone classes to a particular stimulus. In fact, it is the antagonistic center-surround receptive field that allows this comparison, as the center and surround components each receive input from a different class of cone photoreceptor; these neurons are therefore referred to as *color-opponent.* P cells are red-green (L-M) opponent and are the most numerous in the monkey retina, accounting for the entire foveal representation.

The M cells project to the magnocellular (large cell) layers of the LGN, and are found exclusively outside of the fovea. M cells are fewer in number than the P cells, approximately 100,000. M cells have larger dendritic fields than P cells, and because of this are also referred to as parasol cells. The larger dendritic field provides a substrate for greater convergence of bipolar cells onto M cells, which creates a larger receptive field for these cells compared to P cells. M cells typically have larger-diameter axons than P cells, and thus conduct action potentials faster.

The M cells are particularly sensitive to luminance contrast in comparison to P cells, which is somewhat surprising, considering the predominance of P cells in the foveal representation and their small receptive fields. M cells can respond reliably to stimuli with luminance contrasts under 5%, whereas P cells rarely respond reliably to stimuli under 10% contrast. This is an important finding because it has been shown in discrimination experiments that monkeys can detect stimulus changes of under 1% contrast change, meaning that the M channel must be recruited for the discrimination of low-contrast stimuli. Additionally, the M channel has been shown to be the primary source of input to circuits involved in the processing of motion information.

The K Channel

A third class of retinal ganglion cells projects to the thin intercalated layers (or koniocellular "K" layers) of the LGN that lie between the parvocellular and magnocellular layers. Less is known about the physiology of koniocellular-projecting cells, with the exception that many are blue-yellow (S-L/M) opponent, receiving input from the L, M, and S cones. Although there are approximately 100,000 koniocellular cells in the retina and LGN, similar in number to M cells, these cells are extremely small, and the space occupied by them is much less than that occupied by the P or M cells.

Summary of the Response Properties of the P and M Pathways

In summary, there are 5 principle characteristics of the P and M pathways that distinguish them as functionally separate parallel pathways that carry different information regarding the visual environment.

1. P cells are color-sensitive; M-cells are not.
2. M cells respond more strongly to low-contrast, luminance-modulated stimuli than P cells.
3. At a given retinal eccentricity, P cells have smaller receptive fields than M cells.
4. P cell responses are typically sustained; M cell responses are typically transient.
5. M cell axons are larger and conduct action potentials faster than P cell axons.

The Pathway to Cortex

Information leaving the eye is relayed to the visual cortex via neurons in the lateral geniculate nucleus (LGN) of the thalamus. Both the separation of functional streams (parallel pathways) and the visuospatial organization of the retina are maintained in the projections of retinal ganglion cells to the LGN and in the projections of LGN cells to the primary visual cortex.

Retinotopy and the Geometry of Visual Representation

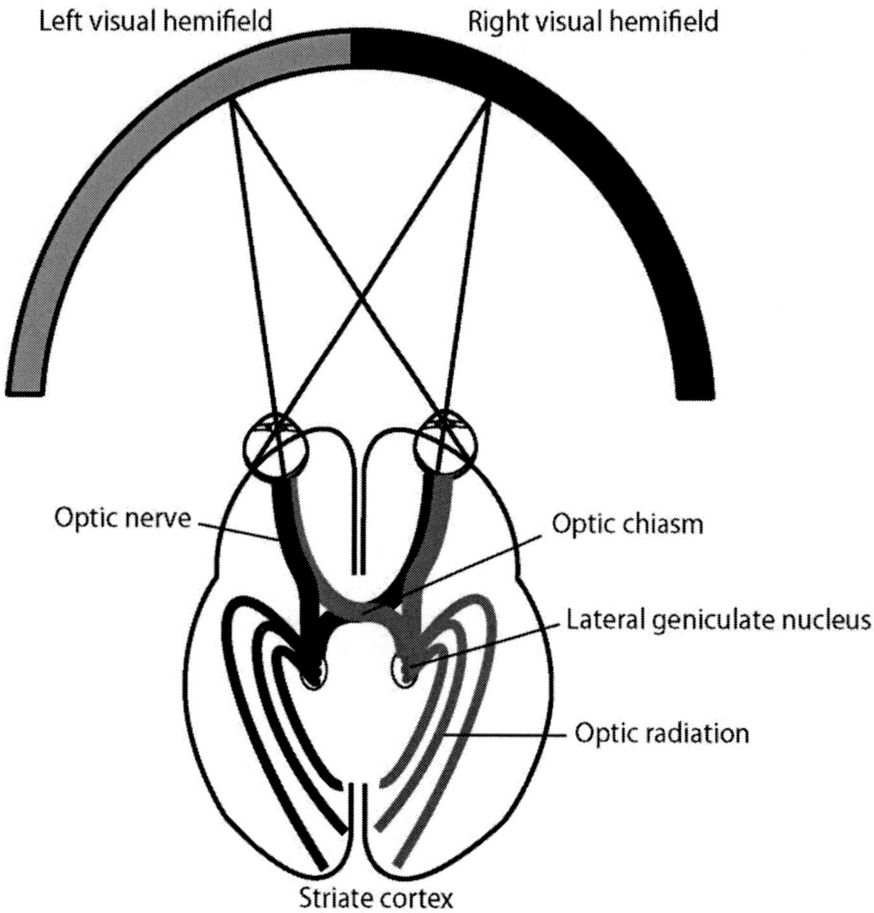

Figure 5. The retino-geniculo-cortical pathway and the visual field representation. Right and left visual hemifields are denoted in black and gray, respectively. Axons from the nasal retina of each eye decussate at the optic chiasm in their projection to the LGN, whereas axons from the temporal retina remain ipsilateral. From this organization, the LGN and V1 on each side of the brain processes visual information from the contralateral visual hemifield.

Due to the optics of the eye, stimuli located in adjacent regions of visual space are projected to adjacent regions of the retina. This neural mapping of space is referred to as *retinotopy* and is essential to our understanding of how the visual scene is initially represented in the brain. The organization of retinal inputs to the two LGN preserves this retinotopy with the exception that each LGN only contains a map of the contralateral visual hemifield rather than the entire visual field (figure 5). Because the visual fields of the two eyes are largely overlapping, this neural representation requires a reorganization of the projections from the eyes to the LGN, such that axons from the temporal retina of each eye project to the ipsilateral LGN while axons from each nasal retina of each eye project to the contralateral LGN. The site where retinal ganglion cell axons cross from one side to the other is called the *optic chiasm* and is located on the ventral surface of the brain, midway between the eyes and the LGN. Projections from the LGN to the primary visual cortex are strictly ipsilateral. As a consequence, neurons in the left primary visual cortex are excited by stimuli in the right visual hemifield, while neurons in the right primary visual cortex are excited by stimuli in the left visual hemifield.

The Lateral Geniculate Nucleus

The LGN is composed of distinctly delimited layers that preserve the ocular, functional, and spatial properties of the retinal ganglion cells that provide input. In most primates, including macaque monkeys and humans, the LGN consists of six primary layers (see figure 7b). The two ventral layers, layers 1 and 2, are the magnocellular layers and the four dorsal layers, layers 3-6, are the parvocellular layers. The contralateral eye provides input to layers 1, 4, and 6 receive input from the contralateral eye and the ipsilateral eye provides input to layers 2, 3, and 5. Between and below each of the primary layers are the intercalated layers that contain koniocellular neurons. Retinal input to the intercalated layers is also eye-specific input, following the ocular specificity of the adjacent and dorsal primary layer.

There are two major classes of neurons within each layer of the LGN: relay neurons and interneurons. Relay neurons use glutamate for communication (excitatory neurotransmitter) and send long-distance axons that provide visual input to the primary visual cortex on the same side of the brain. In contrast, interneurons use GABA for communication (inhibitory neurotransmitter) and have local axons that target nearby neurons in the LGN and are thought to participate in circuitry that influences the timing of the relay cell responses. Interestingly, each relay neuron in the macaque monkey is thought to receive input from only a single retinal ganglion cell. As a consequence, the receptive field properties of relay neurons are very similar to those of their retinal inputs. Relay neurons also provide input to the *reticular nucleus*, a shell surrounding the LGN that is composed entirely of inhibitory neurons. The inhibitory neurons of the reticular nucleus send axons back to the LGN, thus forming a disynaptic, inhibitory loop with the LGN. Such recurrence is a theme in the visual system and, similarly, a massive number of excitatory, feedback projections complete a loop between the primary visual cortex and the LGN.

The Primary Visual Cortex: Gross Anatomy and Visual Representation

In macaque monkeys and other primates with cortical folding, the primary visual cortex (V1) occupies a region that spans some very complex cortical topography and includes a fairly large area that is hidden from view without cutting and unfolding the cortex. Looking at the back of the brain, one can see the *occipital lobe,* the pole of the brain that is responsible for primary visual processing. The most caudal and dorsal region of the occipital lobe is a portion of V1 called the *operculum*, and includes the representation of visual space corresponding to the fovea. Without opening the brain, this is the only portion of V1 that is visible (figure 6a). If one splits the brain down the middle and exposes the interior surface of the hemispheres, more of V1 becomes visible, on the regions of cortex surrounding the *calcarine sulcus,* around which V1 is further involuted (figure 6b). One of the most important aspects of the anatomy of visual cortex is its laminar organization. V1 is composed of six principal layers, numbered 1-6 from the pial surface to the white matter. V1 is also referred to as *striate cortex* because of a heavy band of myelinated axons within layer 4, the 'stria of Gennari', that can be discerned with the naked eye and clearly demarcates the boundaries of V1. Visual areas downstream of V1 are consequently referred to as *extrastriate cortex.*

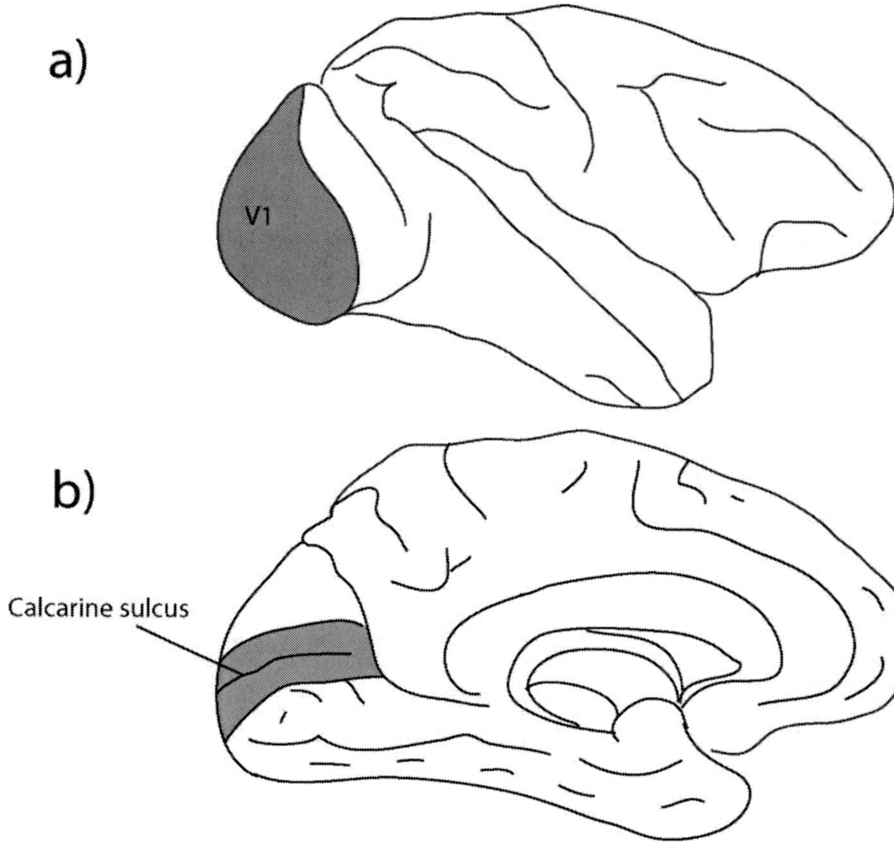

Figure 6. Location of Primary visual cortex (V1) on the cortical surface of the macaque. a) View of the right hemisphere of macaque brain with visible region of V1 shaded. b) View of the medial surface of the left brain hemisphere showing the calcarine sulcus and visible portion of V1.

Streams of Information in V1

The 3 parallel-processing streams established in the circuitry of the retina and maintained in the laminar organization the LGN—the magnocellular, parvocellular, and koniocellular streams—are preserved in the organization of specific connections between the LGN and V1. In the geniculocortical pathway, magnocellular and parvocellular LGN axons terminate selectively in layers 4Cα and 4Cβ, respectively (figure 7). Koniocellular axons bypass these layers and terminate in layers 2 and 3, in delimited regions that can be identified by histological staining of the tissue. When sections of V1 are stained for *cytochrome oxidase*, dark regions in layers 2 and 3, referred to as 'blobs', correspond to the projection patterns of the koniocellular axons. Neurons in the blobs and interblobs are the subject of much discussion and debate, as evidence indicates there may be a differential distribution of color-selective and orientation-selective neurons in these compartments.

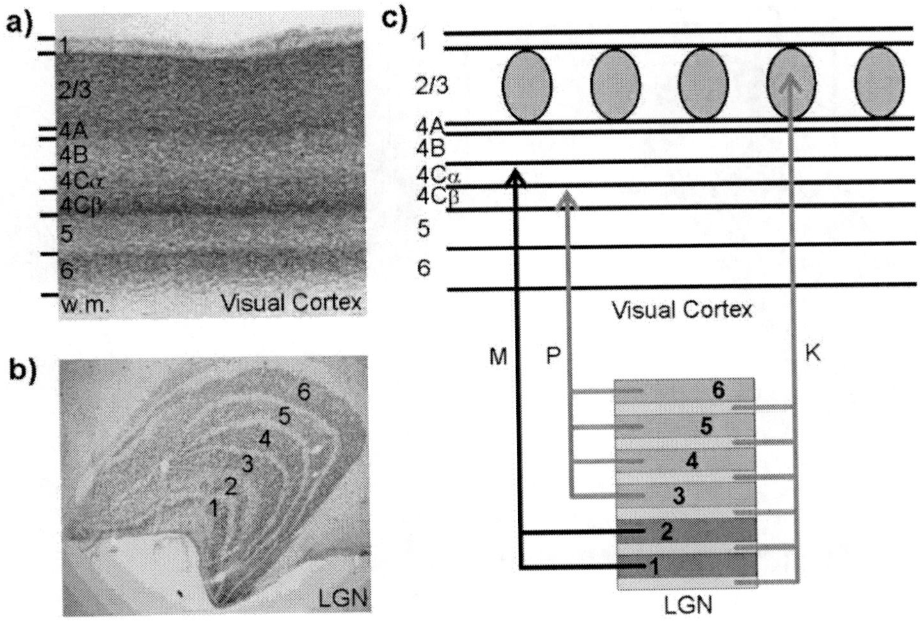

Figure 7. Anatomy of feedforward and feedback connections between the LGN and visual cortex (V1). a and b) Nissl-stained sections of V1 and the LGN. Both structures are highly laminated. LGN layers 1 and 2 are the magnocellular layers; layers 3, 4, 5 and 6 are the parvocellular layers. c) Organization of connections between the LGN and V1. Magnocellular LGN axons terminate in layers 4Cα, parvocellular LGN axons terminate in layers 4Cβ, and koniocellular axons terminate in the blobs of layers 2 and 3.

Simple Cells and Complex Cells in V1: Response Characteristics

As mentioned above, there is little or no convergence in the pathway from retina to LGN—a property that confers the response characteristics of individual retinal ganglion cells onto their targets. In contrast, there is heterogeneity in the amount of convergence occurring in the geniculocortical pathway. Although more work is needed to determine the details governing geniculocortical connectivity in the macaque monkey, existing evidence indicates convergence is more widespread among magnocellular inputs to neurons in layer 4Cα neurons than parvocellular inputs to neurons in layer 4Cβ. Convergence has important implications on the response properties of target neurons, as new receptive fields can be formed by combining inputs with distinct properties. Hubel and Wiesel (1962) provided the first model describing how the convergence of LGN inputs could underlie the emergence of a new receptive field in the cortex (figure 8). In their model, LGN neurons with receptive fields of the same sign (on or off) and located along a line in visual space provide convergent input to a target cortical neuron. As a consequence of this organization, postsynaptic neurons (termed "simple cells"), would respond selectively to a stimulus elongated along a particular axis of orientation, while presynaptic neurons would respond equally well to stimuli of all orientations.

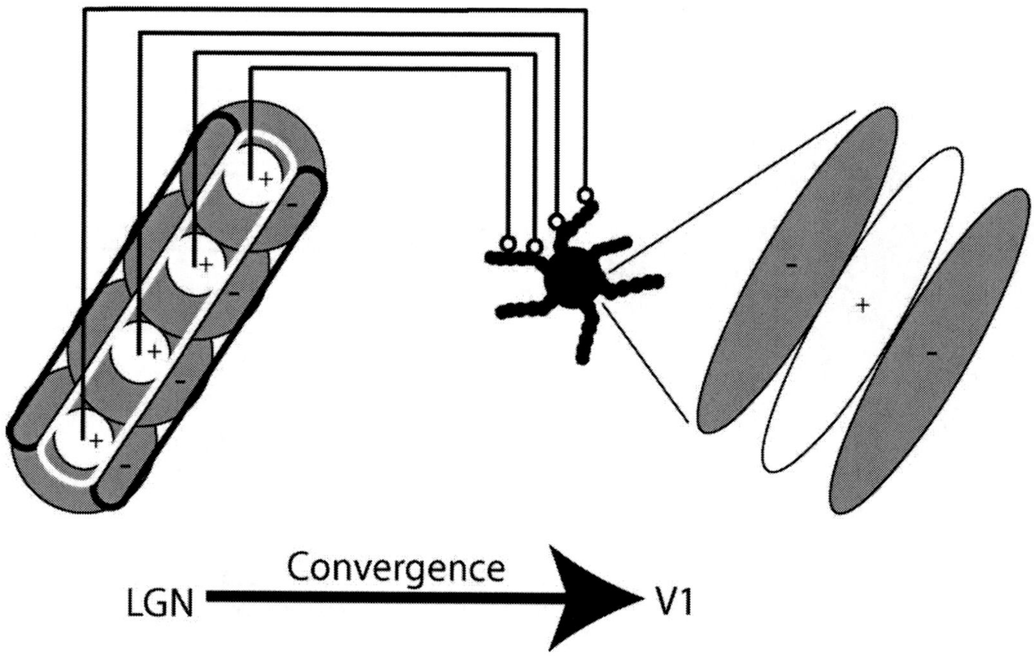

Figure 8. LGN cell convergence creates oriented V1 RFs. LGN cells, denoted above by their receptive fields, provide input to a spiny stellate neuron in V1; the convergent input of cells with offset receptive fields results in a conglomerate receptive field with an elongated axis. Following Hubel and Wiesel (1962).

In addition to simple cells, Hubel and Wiesel described a second class of neurons they called "complex cells" (Hubel and Wiesel, 1962, 1968). The distinction between simple and complex cells is fundamental to understanding the early visual system, although the distinction has necessarily become more complicated as decades of research have progressed. The classification of these cells is determined by their response properties, and exactly how this should be done has garnered some debate. Figure 8 shows a schematic of an ideal simple cell receptive field; these receptive fields are composed of distinct *on* and *off* subregions, corresponding to the receptive field characteristics of the LGN relay cells that define their input. These cells respond predictably to light increments or decrements occurring in the corresponding regions of their receptive fields. Complex cell receptive fields are much harder to characterize. These cells do not exhibit distinct *on* and *off* subregions—they respond to both increments and decrements of light at all locations in their receptive fields. Hubel and Wiesel hypothesized that complex cells receive their inputs from heterogeneous groups of simple cells, creating these response profiles. Complex cells respond best to stimuli of a particular orientation, and thus it is hypothesized that they receive convergent input from simple cells sharing the same orientation preference, but differing in the locations of their on and off subregions.

Columnar Organization in V1: Ocular Dominance and Tuning

Most neurons in V1 exhibit an ocular dominance, meaning that although they will respond to stimuli presented to either eye, they respond more strongly to stimuli presented to a preferred eye. With this in mind, there is a range in the ocular bias of V1 neurons, with some neurons being completely monocular, some neurons being almost completely binocular, and others falling along a gradient between these extremes. The orientation tuning profiles of V1 neurons is similar, with the extremes being cells that exhibit almost perfect orientation tuning, and cells with very broad orientation tuning.

One of the most interesting aspects of V1 is the way that it is organized with regard to the functional specificity of the neurons that compose it. Neurons in V1 with similar response properties reside near each other, forming sub-populations of cells that represent particular aspects of the visual environment. Within the retinotopic map of V1, there is further functional organization of neurons sharing similar ocular preference and orientation tuning. These populations form columns through the layers of the cortex, such that cells with similar response properties are delimited across the surface of the cortex, but span the entire depth of the 6 layers. Hubel and Wiesel found that when an electrode was advanced through the cortex perpendicular to the cortical surface, neurons with similar response properties were encountered throughout the entire penetration. Conversely, when an electrode was advanced through the cortex more tangentially to the surface, the response properties of the neurons encountered changed gradually as the electrode was advanced. These functional columns are referred to as *ocular dominance columns* and *orientation columns*.

Beyond V1: The Extrastriate Areas

More than half of the macaque cortex is devoted to visual processing. While V1 represents the first stage processing in the cortical hierarchy, there are more than 30 functionally and anatomically defined areas downstream of V1, residing in large portions of the parietal and temporal lobes. Research investigating the precise forms of processing and transformations in these particular areas is a fiercely active field of vision science today, and will likely remain so for many years. In the following section, we present the foundational information necessary to understand the basic organization and function of the extrastiate cortical areas.

The three segregated streams of input to V1 is both mixed and maintained within the intrinsic circuits of V1. Although the details of the mixing and segregation of circuits remains to be determined, the outputs from V1 to V2 show a functional specificity. Namely, neurons in the blobs and interblobs of layers 2 and 3 project to two compartments in V2, the thin and pale stripes (also revealed in cytochrome-oxidase stained tissue). In contrast, neurons in layer 4B of V1 project to neurons in the thick stripes of V2. The compartmental organization of V2 is important, as neurons in the thick stripes provide input to cortical areas in the dorsal stream, while neurons in the thin and pale stripes provide input to cortical areas in the ventral stream (described below).

The large-scale, functional organization of the extrastriate areas was elegantly demonstrated by experiments performed by Ungerleider and Mishkin (1982). In these experiments (Figure 9), restricted lesions to visual areas in either the temporal lobe or the parietal lobe led to distinct behavioral deficits. Temporal lobe lesions caused deficits in an object recognition task but not in a spatial discrimination task; conversely, parietal lesions caused deficits in a spatial discrimination task but not in an object recognition task. Together with physiological evidence supporting such functional specificity, these results revealed the presence of parallel processing streams in the extrastriate cortices: a *ventral stream* involved in the processing of object information, and a *dorsal stream* involved in the processing of actions and spatial tasks. One of the most interesting aspects of this separation is that the ventral stream, V1-V2-V3-V4-IT, is primarily fed by the parvocellular and koniocellular stream, while the dorsal stream, V1-V2-MT-MST, is primarily fed by the magnocellular stream. Thus, the functional parallel streams established in the retina are maintained (although not strictly so) through higher cortical processing.

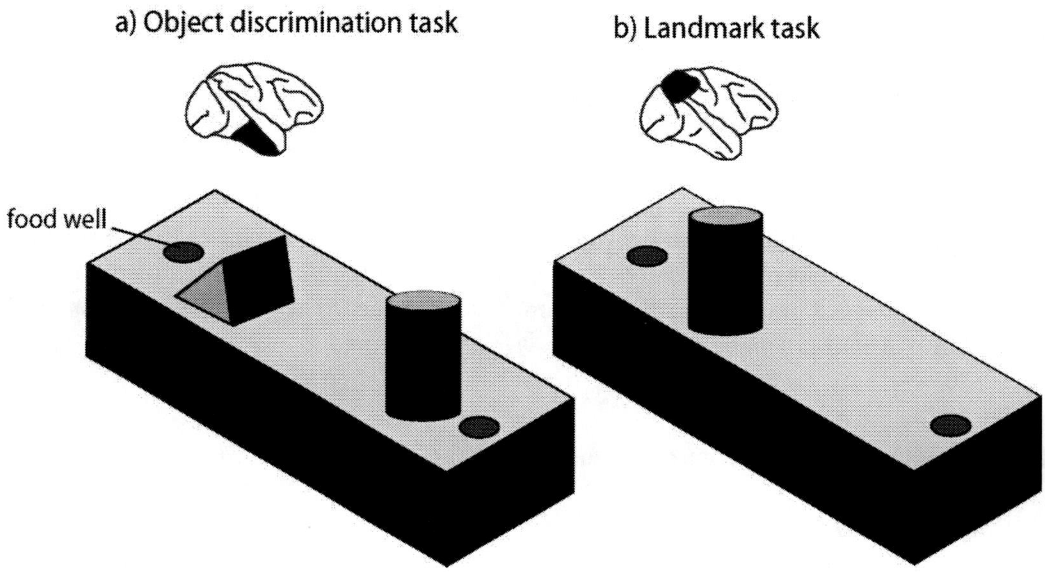

Figure 9. Dissociation of dorsal stream and ventral stream in macaque monkey. Ungerleider and Mishkin (1982), trained monkeys to reach for a food well near a particular object shape (a) or near a landmark (b) to obtain a food reward. Monkeys with temporal lesion (see under 'a') showed a deficit in the object discrimination task, but not the landmark task; conversely, monkeys with parietal lesions (under 'b') showed deficits in the landmark task, but not the object discrimination task.

The physiology of the dorsal stream is best introduced by considering the response properties of neurons in area MT (also called V5). Neurons in area MT respond selectively to the motion of a visual stimulus in a particular direction. The receptive fields of MT neurons are much larger than those found in V1; consequently, motion information is integrated over broad regions. Along with neurons in area MST (also in the dorsal stream), MT neurons appear well-suited for processing visual stimuli resulting self-movement.

Neurons in the fusiform face area of the ventral stream provide a nice example of visual processing specific to objects. In 1981, Bruce et al. demonstrated that some neurons in the temporal cortex responded preferentially to faces over other categories of visual stimuli. Face selective cells display of range of response properties, with some cells strongly tuned to respond to complete faces, and other cells tuned to particular aspects of faces such as eye separation or direction of gaze. More recent studies have suggested that these populations of neurons are in fact organized in patches on the superior temporal sulcus (Tsao et al., 2003, 2006). These examples reinforce several overall points in the organization of the visual system: functional separation of processing streams, a gradual specialization of receptive field properties at subsequent stages in the visual hierarchy, and organization of cell populations into functional maps within the borders of defined visual areas.

REFERENCES

Bruce, C., Desimone, R., and Gross, C.G. (1981). Visual properties of neurons in a polysensory area in superior temporal sulcus of the macaque. *J. Neurophysiol.* 46:369-384.

Hartline, H.K. (1938). The response of single optic nerve fibers of the vertebrate eye to illumination of the retina. Am. J. Physiol. 121:400--415.

Hubel, D.H., and Wiesel, T.N. (1968). Receptive fields and functional architecture of monkey striate cortex. J. Physiol. 195:215-243.

Hubel, D.H., and Wiesel, T.N. (1962) Receptive fields, binocular interaction and functional architecture in the cat's visual cortex. *J. Physiol.* 160:106–154.

Tsao, D.Y., Freiwald, W.A., Knutsen, T.A., Mandeville, J.B., and Tootell, R.B.H. (2003). The representation of faces and objects in Macaque Cerebral Cortex. *Nat. Neurosci.* 6:989-995.

Tsao, D.Y., Freiwald, W.A., Tootell, R.B.H.,and Livingstone, M.S.L. (2006). A cortical region consisting entirely of face cells. *Science* 311:670-674.

Ungerleider, L.G., and Mishkin, M. (1982). Two cortical visual systems. *In Analysis of visual behavior* (ed. D. J. Ingle, M. A. Goodale & R. J. W. Mansfield). Cambridge, MA: MIT Press.

Wandell, B.A. (1995). *Foundations of Vision*. Sinaur Associates, Inc. Sunderland, MA.

INDEX

2

A

B

C

D

E

F

G

H

I

N

Q

R

S

T